Modern Digital Communications

Modern Digital Communications

Stephen Holland

Larsen & Keller
www.larsen-keller.com

Modern Digital Communications
Stephen Holland
ISBN: 978-1-64172-660-3 (Hardback)

☰ Larsen & Keller

Published by Larsen and Keller Education,
5 Penn Plaza,
19th Floor,
New York, NY 10001, USA

Cataloging-in-Publication Data

Modern digital communications / Stephen Holland.
    p. cm.
Includes bibliographical references and index.
ISBN 978-1-64172-660-3
1. Digital communications. 2. Telecommunication. 3. Digital electronics.
4. Digital media. 5. Signal processing--Digital techniques. I. Holland, Stephen.
TK5103.7 .M63 2022
621.382--dc23

For more information regarding Larsen and Keller Education and its products, please visit the publisher's website www.larsen-keller.com

Table of Contents

Preface   VII

Chapter 1 Introduction   1
- Digital Communication   1
- Quantization   10
- Digital Modulation   35
- Digital Signal   51
- Characteristics of a Digital Signal   55
- Nyquist–Shannon Sampling Theorem   56

Chapter 2 Pulse Modulation   66
- Pulse Position Modulation   71
- Pulse Code Modulation   73
- Differential Pulse Code Modulation   82
- Delta Modulation   85
- Time Division Multiplexing   88
- Frequency Division Multiplexing   91
- Carrier Systems   94

Chapter 3 Digital Carrier Modulation Techniques   104
- Digital Communication System   104
- Amplitude Shift Keying   108
- Frequency Shift Keying   111
- Phase Shift Keying   115
- M-Ary Encoding   132

Chapter 4 Multiple Access Techniques   135
- Frequency Division Multiple Access   138
- Time Division Multiple Access   139
- Code Division Multiple Access   142
- Space Division Multiple Access   150

Chapter 5	Modern Media for Communication	151
▪	Mobile Communication	154
▪	Satellite Communication	166

Chapter 6	Diverse Aspects of Digital Communication	173
▪	Bit Rate	173
▪	Baud Rate	178
▪	Error Detection and Correction Code	186
▪	Bandwidth	202
▪	Noise	205
▪	Signal to Noise Ratio	209
▪	Modulation Error Ratio	215
▪	Shannon–Hartley Theorem	216
▪	Transmitter	220
▪	Receiver	223

Permissions

Index

Preface

This book aims to help a broader range of students by exploring a wide variety of significant topics related to this discipline. It will help students in achieving a higher level of understanding of the subject and excel in their respective fields. This book would not have been possible without the unwavered support of my senior professors who took out the time to provide me feedback and help me with the process. I would also like to thank my family for their patience and support.

The transfer of data from one point to another is known as digital communication. Optical fibers, storage media, computer buses, wireless communication channels and copper wires are a few types of channels used in digital communications. The data in digital communication is represented as an electrical voltage, microwave, radiowave and infrared signal. Synchronous and asynchronous transmissions are the two types of data communication. The transmission speeds at both the sending and receiving end of the transmission are synchronized in synchronous transmission by using clock signals. Asynchronous serial communication uses start and stop bits to signify the beginning and end of a transmission. Digital communications is an upcoming field of science that has undergone rapid development over the past few decades. Some of the diverse topics covered herein address the varied concepts and applications that fall under this category. Scientists and students actively engaged in this field will find this book full of crucial and unexplored concepts.

A brief overview of the book contents is provided below:

Chapter – Introduction

The transmission of data over different communication channels such as point to point or point to multipoint is known as digital communication. This is an introductory chapter which will introduce briefly all the significant aspects of digital communication such as quantization, digital modulation, digital signal, etc.

Chapter – Pulse Modulation

Pulse modulation is a technique where signal is transmitted along with the information by pulses. It is categorized into pulse-width modulation, pulse position modulation, pulse code modulation and delta modulation. This chapter discusses in detail these different methods and techniques related to pulse modulation.

Chapter – Digital Carrier Modulation Techniques

The modulation techniques which use discrete signals to modulate a carrier wave are known as digital carrier modulation. This chapter has been carefully written to provide an easy understanding of the various types of digital carrier modulation techniques such as amplitude shift keying, frequency shift keying, phase shift keying, etc.

Chapter – Multiple Access Techniques

The techniques that let multiple mobile users share the allotted spectrum in the most effective manner for improving the overall access capacity are known as multiple access techniques. A few of such techniques are frequency division multiple access, time division multiple access and space division multiple access. The topics elaborated in this chapter will help in gaining a better perspective about these multiple access techniques.

Chapter – Modern Media for Communication

There are numerous tools which are used for communication such as social media, SMS text messaging, email marketing, direct email, blogging, voice calling, video chat, video marketing, live web chat and virtual reality. This chapter has been carefully written to provide an easy understanding of these methods of modern communication.

Chapter – Diverse Aspects of Digital Communication

There are numerous aspects of digital communications such as bit rate, baud rate error detection and correction code, bandwidth, signal to noise ratio, modulation error ratio, Shannon-Hartley theorem, etc. This chapter closely examines these aspects of digital communications to provide an extensive understanding of the subject.

Stephen Holland

1

Introduction

The transmission of data over different communication channels such as point to point or point to multipoint is known as digital communication. This is an introductory chapter which will introduce briefly all the significant aspects of digital communication such as quantization, digital modulation, digital signal, etc.

DIGITAL COMMUNICATION

Digital communications is any exchange of data that transmits the data in a digital form. For example, communications done over the Internet is a form of digital communication.

The communication that occurs in our day-to-day life is in the form of signals. These signals, such as sound signals, generally, are analog in nature. When the communication needs to be established over a distance, then the analog signals are sent through wire, using different techniques for effective transmission.

Necessity of Digitization

The conventional methods of communication used analog signals for long distance communications, which suffer from many losses such as distortion, interference, and other losses including security breach.

In order to overcome these problems, the signals are digitized using different techniques. The digitized signals allow the communication to be more clear and accurate without losses.

Advantages of Digital Communication

As the signals are digitized, there are many advantages of digital communication over analog communication, such as:

- The effect of distortion, noise, and interference is much less in digital signals as they are less affected.

- Digital circuits are more reliable.

- Digital circuits are easy to design and cheaper than analog circuits.

- The hardware implementation in digital circuits, is more flexible than analog.

- The occurrence of cross-talk is very rare in digital communication.

- The signal is un-altered as the pulse needs a high disturbance to alter its properties, which is very difficult.

- Signal processing functions such as encryption and compression are employed in digital circuits to maintain the secrecy of the information.

- The probability of error occurrence is reduced by employing error detecting and error correcting codes.

- Spread spectrum technique is used to avoid signal jamming.

- Combining digital signals using Time Division Multiplexing (TDM) is easier than combining analog signals using Frequency Division Multiplexing (FDM).

- The configuring process of digital signals is easier than analog signals.

- Digital signals can be saved and retrieved more conveniently than analog signals.

- Many of the digital circuits have almost common encoding techniques and hence similar devices can be used for a number of purposes.

- The capacity of the channel is effectively utilized by digital signals.

Elements of Digital Communication

The elements which form a digital communication system is represented by the following block diagram for the ease of understanding.

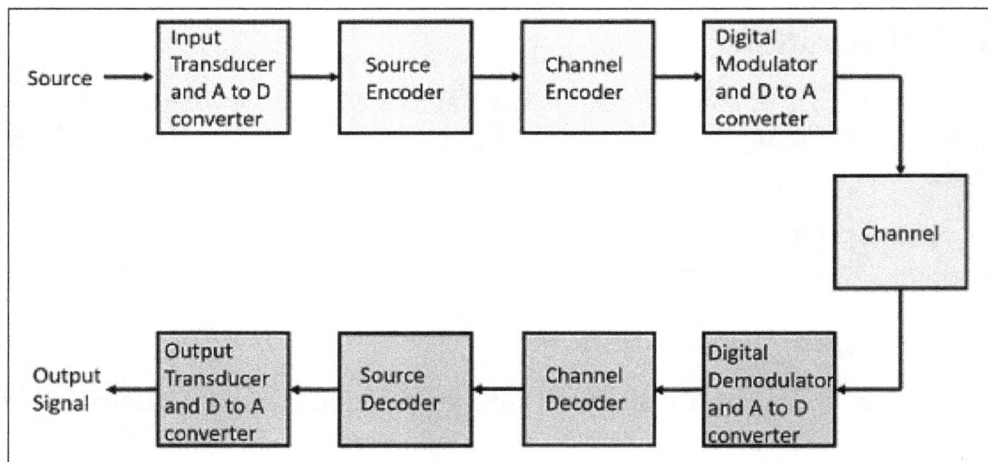

Basic Elements of Digital Communication System.

Following are the sections of the digital communication system.

Source

The source can be an analog signal. Example: A Sound signal.

Input Transducer

This is a transducer which takes a physical input and converts it to an electrical signal (Example: microphone). This block also consists of an analog to digital converter where a digital signal is needed for further processes.

A digital signal is generally represented by a binary sequence.

Source Encoder

The source encoder compresses the data into minimum number of bits. This process helps in effective utilization of the bandwith. It removes the redundant bits (unnecessary excess bits, i.e., zeroes).

Channel Encoder

The channel encoder, does the coding for error correction. During the transmission of the signal, due to the noise in the channel, the signal may get altered and hence to avoid this, the channel encoder adds some redundant bits to the transmitted data. These are the error correcting bits.

Digital Modulator

The signal to be transmitted is modulated here by a carrier. The signal is also converted to analog from the digital sequence, in order to make it travel through the channel or medium.

Channel

The channel or a medium, allows the analog signal to transmit from the transmitter end to the receiver end.

Digital Demodulator

This is the first step at the receiver end. The received signal is demodulated as well as converted again from analog to digital. The signal gets reconstructed here.

Channel Decoder

The channel decoder, after detecting the sequence, does some error corrections. The distortions which might occur during the transmission, are corrected by adding some redundant bits. This addition of bits helps in the complete recovery of the original signal.

Source Decoder

The resultant signal is once again digitized by sampling and quantizing so that the pure digital output is obtained without the loss of information. The source decoder recreates the source output.

Output Transducer

This is the last block which converts the signal into the original physical form, which was at the input of the transmitter. It converts the electrical signal into physical output (Example: loud speaker).

Output Signal

This is the output which is produced after the whole process. Example – The sound signal received.

This unit has dealt with the introduction, the digitization of signals, the advantages and the elements of digital communications.

Modulation

Modulation is the process of varying one or more parameters of a carrier signal in accordance with the instantaneous values of the message signal.

The message signal is the signal which is being transmitted for communication and the carrier signal is a high frequency signal which has no data, but is used for long distance transmission.

There are many modulation techniques, which are classified according to the type of modulation employed. Of them all, the digital modulation technique used is Pulse Code Modulation (PCM).

A signal is pulse code modulated to convert its analog information into a binary sequence, i.e., 1s and 0s. The output of a PCM will resemble a binary sequence. The following figure shows an example of PCM output with respect to instantaneous values of a given sine wave.

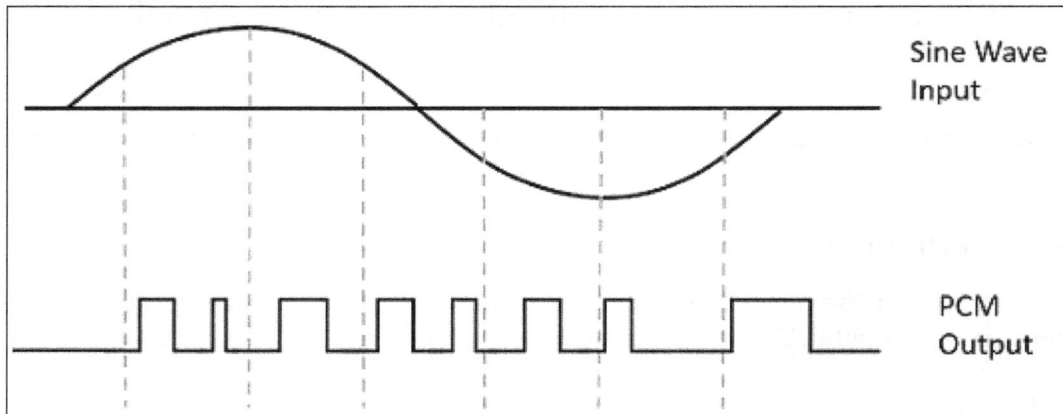

Instead of a pulse train, PCM produces a series of numbers or digits, and hence this process is called as digital. Each one of these digits, though in binary code, represent the approximate amplitude of the signal sample at that instant.

In Pulse Code Modulation, the message signal is represented by a sequence of coded pulses. This message signal is achieved by representing the signal in discrete form in both time and amplitude.

Basic Elements of PCM

The transmitter section of a Pulse Code Modulator circuit consists of Sampling, Quantizing and Encoding, which are performed in the analog-to-digital converter section. The low pass filter prior to sampling prevents aliasing of the message signal.

The basic operations in the receiver section are regeneration of impaired signals, decoding, and reconstruction of the quantized pulse train. Following is the block diagram of PCM which represents the basic elements of both the transmitter and the receiver sections.

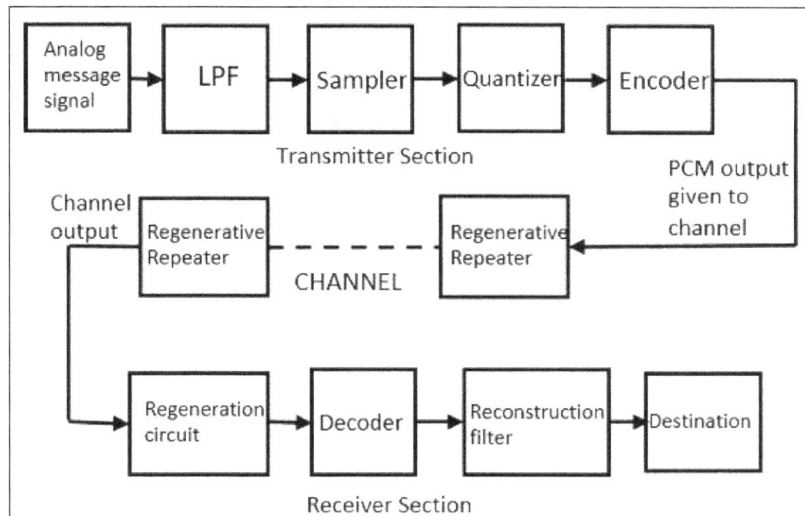

Low Pass Filter

This filter eliminates the high frequency components present in the input analog signal which is greater than the highest frequency of the message signal, to avoid aliasing of the message signal.

Sampler

This is the technique which helps to collect the sample data at instantaneous values of message signal, so as to reconstruct the original signal. The sampling rate must be greater than twice the highest frequency component W of the message signal, in accordance with the sampling theorem.

Quantizer

Quantizing is a process of reducing the excessive bits and confining the data. The sampled output when given to Quantizer, reduces the redundant bits and compresses the value.

Encoder

The digitization of analog signal is done by the encoder. It designates each quantized level by a binary code. The sampling done here is the sample-and-hold process. These three sections (LPF, Sampler, and Quantizer) will act as an analog to digital converter. Encoding minimizes the bandwith used.

Regenerative Repeater

This section increases the signal strength. The output of the channel also has one regenerative repeater circuit, to compensate the signal loss and reconstruct the signal, and also to increase its strength.

Decoder

The decoder circuit decodes the pulse coded waveform to reproduce the original signal. This circuit acts as the demodulator.

Reconstruction Filter

After the digital-to-analog conversion is done by the regenerative circuit and the decoder, a low-pass filter is employed, called as the reconstruction filter to get back the original signal.

Hence, the Pulse Code Modulator circuit digitizes the given analog signal, codes it and samples it, and then transmits it in an analog form. This whole process is repeated in a reverse pattern to obtain the original signal.

Sampling

Sampling is defined as, "The process of measuring the instantaneous values of continuous-time signal in a discrete form".

Sample is a piece of data taken from the whole data which is continuous in the time domain.

When a source generates an analog signal and if that has to be digitized, having 1s and 0s i.e., High or Low, the signal has to be discretized in time. This discretization of analog signal is called as Sampling.

The following figure indicates a continuous-time signal x (t) and a sampled signal x_s (t). When x (t) is multiplied by a periodic impulse train, the sampled signal x_s (t) is obtained.

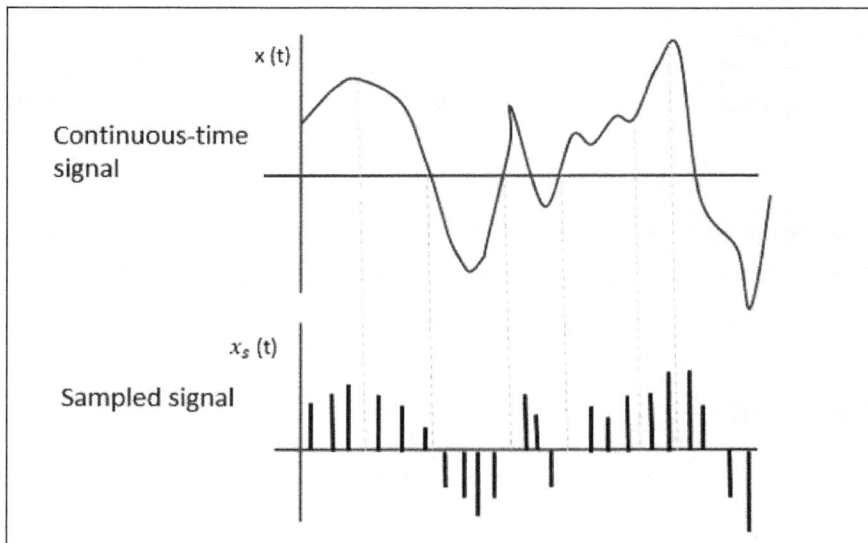

Sampling Rate

To discretize the signals, the gap between the samples should be fixed. That gap can be termed as a sampling period Ts.

Sampling Frequency $= \dfrac{1}{T_S} = f_s$

Where,

- T_s is the sampling time,

- f_s is the sampling frequency or the sampling rate.

Sampling frequency is the reciprocal of the sampling period. This sampling frequency, can be simply called as Sampling rate. The sampling rate denotes the number of samples taken per second, or for a finite set of values.

For an analog signal to be reconstructed from the digitized signal, the sampling rate should be highly considered. The rate of sampling should be such that the data in the message signal should neither be lost nor it should get over-lapped. Hence, a rate was fixed for this, called as Nyquist rate.

Nyquist Rate

Suppose that a signal is band-limited with no frequency components higher than W Hertz. That means, W is the highest frequency. For such a signal, for effective reproduction of the original signal, the sampling rate should be twice the highest frequency.

Which means,

$f_s = 2W$

where,

- f_s is the sampling rate,

- W is the highest frequency.

This rate of sampling is called as Nyquist rate.

A theorem called, Sampling Theorem, was stated on the theory of this Nyquist rate.

Sampling Theorem

The sampling theorem, which is also called as Nyquist theorem, delivers the theory of sufficient sample rate in terms of bandwith for the class of functions that are bandlimited.

The sampling theorem states that, "a signal can be exactly reproduced if it is sampled at the rate fs which is greater than twice the maximum frequency W."

To understand this sampling theorem, let us consider a band-limited signal, i.e., a signal whose value is non-zero between some −W and W Hertz.

Such a signal is represented as $x(f) = 0$ for $|f| > W$.

For the continuous-time signal x (t), the band-limited signal in frequency domain, can be represented as shown in the following figure.

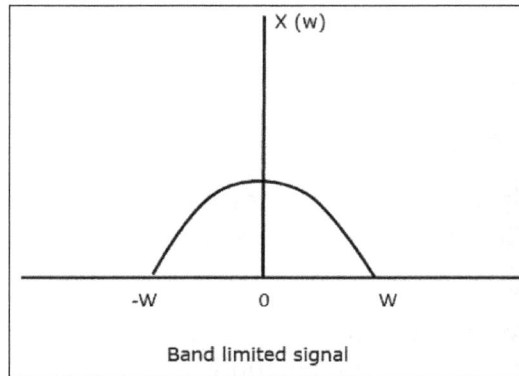

Band limited signal

We need a sampling frequency, a frequency at which there should be no loss of information, even after sampling. For this, we have the Nyquist rate that the sampling frequency should be two times the maximum frequency. It is the critical rate of sampling.

If the signal x(t) is sampled above the Nyquist rate, the original signal can be recovered, and if it is sampled below the Nyquist rate, the signal cannot be recovered.

The following figure explains a signal, if sampled at a higher rate than 2w in the frequency domain.

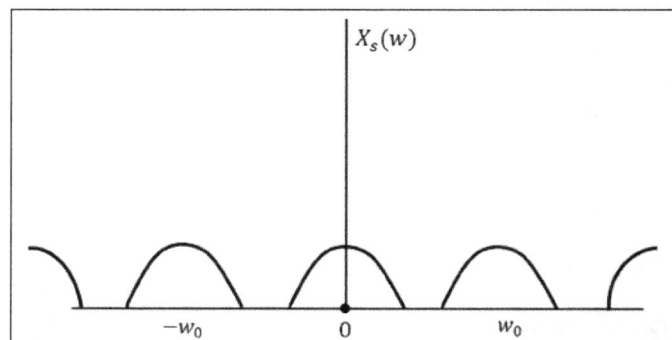

The above figure shows the Fourier transform of a signal xs(t). Here, the information is reproduced without any loss. There is no mixing up and hence recovery is possible.

The Fourier Transform of the signal $x_s(t)$) is:

$$X_s(w) = \frac{1}{T_s} \sum_{n \, \infty}^{\infty} X(w-nw_0)$$

Where T_s = Sampling Period and $w_0 = \dfrac{2\pi}{Ts}$.

Let us see what happens if the sampling rate is equal to twice the highest frequency (2W).

That means,

$$f_s = 2W$$

where,

- f_s is the sampling frequency

- W is the highest frequency

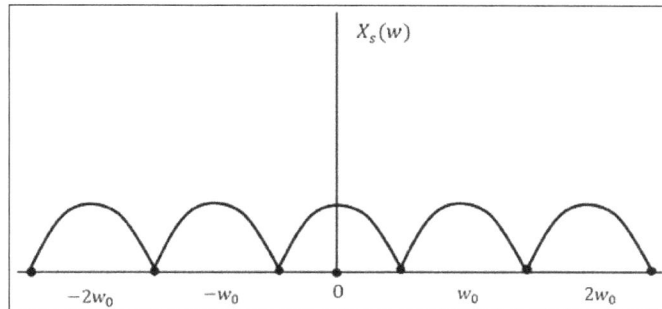

The result will be as shown in the above figure. The information is replaced without any loss. Hence, this is also a good sampling rate.

Now, let us look at the condition,

$$f_2 < 2W$$

The resultant pattern will look like the following figure.

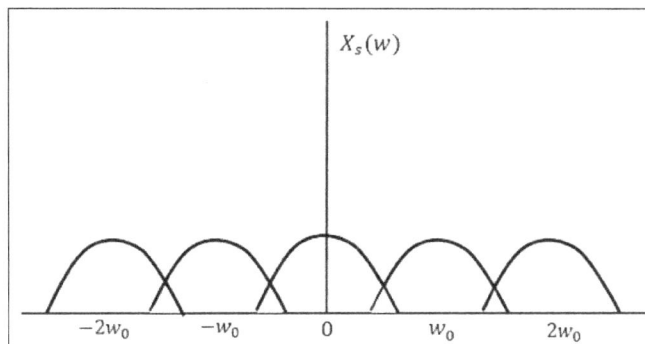

We can observe from the above pattern that the over-lapping of information is done, which leads to mixing up and loss of information. This unwanted phenomenon of over-lapping is called as Aliasing.

Aliasing

Aliasing can be referred to as "the phenomenon of a high-frequency component in the spectrum of a signal, taking on the identity of a low-frequency component in the spectrum of its sampled version."

The corrective measures taken to reduce the effect of Aliasing are:

- In the transmitter section of PCM, a low pass anti-aliasing filter is employed, before the sampler, to eliminate the high frequency components, which are unwanted.

- The signal which is sampled after filtering, is sampled at a rate slightly higher than the Nyquist rate.

This choice of having the sampling rate higher than Nyquist rate, also helps in the easier design of the reconstruction filter at the receiver.

Scope of Fourier Transform

It is generally observed that, we seek the help of Fourier series and Fourier transforms in analyzing the signals and also in proving theorems. It is because:

- The Fourier Transform is the extension of Fourier series for non-periodic signals.

- Fourier transform is a powerful mathematical tool which helps to view the signals in different domains and helps to analyze the signals easily.

- Any signal can be decomposed in terms of sum of sines and cosines using this Fourier transform.

QUANTIZATION

Quantization, in mathematics and digital signal processing, is the process of mapping input values from a large set (often a continuous set) to output values in a (countable) smaller set, often with a finite number of elements. Rounding and truncation are typical examples of quantization processes. Quantization is involved to some degree in nearly all digital signal processing, as the process of representing a signal in digital form ordinarily involves rounding. Quantization also forms the core of essentially all lossy compression algorithms.

The difference between an input value and its quantized value (such as round-off error) is referred to as quantization error. A device or algorithmic function that performs quantization is called a quantizer. An analog-to-digital converter is an example of a quantizer.

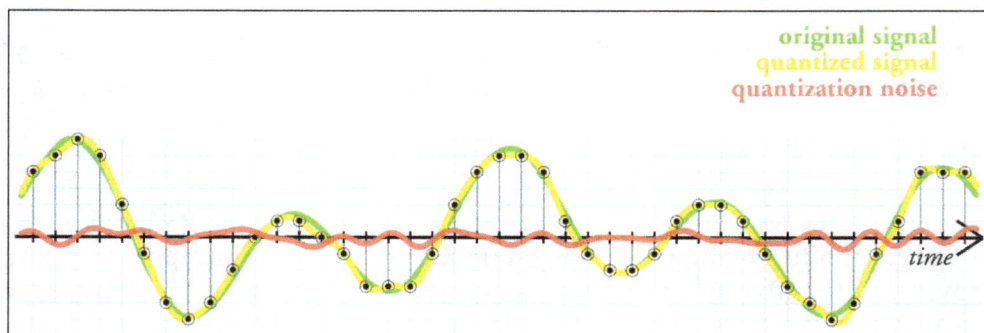

The simplest way to quantize a signal is to choose the digital amplitude value closest to the original analog amplitude. This example shows the original analog signal (green), the quantized signal (black dots), the signal reconstructed from the quantized signal (yellow) and the difference between the original signal and the reconstructed signal (red). The difference between the original signal and the reconstructed signal is the quantization error and, in this simple quantization scheme, is a deterministic function of the input signal.

Mathematical Properties

Because quantization is a many-to-few mapping, it is an inherently non-linear and irreversible process (i.e., because the same output value is shared by multiple input values, it is impossible, in general, to recover the exact input value when given only the output value).

The set of possible input values may be infinitely large, and may possibly be continuous and therefore uncountable (such as the set of all real numbers, or all real numbers within some limited range). The set of possible output values may be finite or countably infinite. The input and output sets involved in quantization can be defined in a rather general way. For example, vector quantization is the application of quantization to multi-dimensional (vector-valued) input data.

Basic Types of Quantization

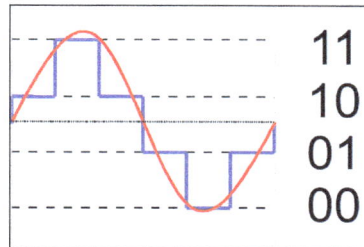

2-bit resolution with four levels of quantization compared to analog.

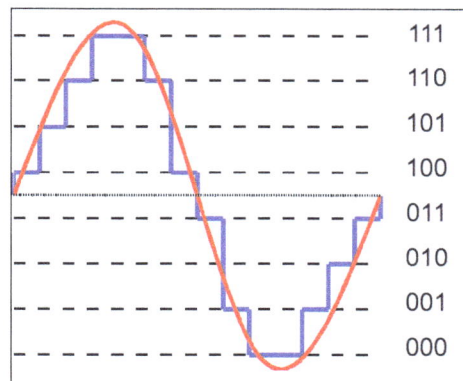

3-bit resolution with eight levels.

Analog-to-digital Converter

An analog-to-digital converter (ADC) can be modeled as two processes: sampling and quantization. Sampling converts a time-varying voltage signal into a discrete-time signal, a sequence of real numbers. Quantization replaces each real number with an approximation from a finite set of discrete values. Most commonly, these discrete values are represented as fixed-point words. Though any number of quantization levels is possible, common word-lengths are 8-bit (256 levels), 16-bit (65,536 levels) and 24-bit (16.8 million levels). Quantizing a sequence of numbers produces a sequence of quantization errors which is sometimes modeled as an additive random signal called quantization noise because of its stochastic behavior. The more levels a quantizer uses, the lower is its quantization noise power.

Rate–distortion Optimization

Rate–distortion optimized quantization is encountered in source coding for lossy data compression algorithms, where the purpose is to manage distortion within the limits of the bit rate supported by a communication channel or storage medium. The analysis of quantization in this context involves studying the amount of data (typically measured in digits or bits or bit *rate*) that is used to represent the output of the quantizer, and studying the loss of precision that is introduced by the quantization process (which is referred to as the distortion).

Rounding Example:

As an example, rounding a real number x to the nearest integer value forms a very basic type of quantizer – a *uniform* one. A typical (*mid-tread*) uniform quantizer with a quantization *step size* equal to some value Δ can be expressed as:

$$Q(x) = \Delta \cdot \left\lfloor \frac{x}{\Delta} + \frac{1}{2} \right\rfloor = \Delta \cdot \text{floor}\left(\frac{x}{\Delta} + \frac{1}{2} \right),$$

where the notation $\lfloor \ \rfloor$ or $(\ )$ depicts the floor function.

The essential property of a quantizer is that it has a countable set of possible output values that has fewer members than the set of possible input values. The members of the set of output values may have integer, rational, or real values. For simple rounding to the nearest integer, the step size is equal to 1. With $\Delta = 1$ or with Δ equal to any other integer value, this quantizer has real-valued inputs and integer-valued outputs.

When the quantization step size (Δ) is small relative to the variation in the signal being quantized, it is relatively simple to show that the mean squared error produced by such a rounding operation will be approximately $\Delta^2 / 12$. Mean squared error is also called the quantization *noise power*. Adding one bit to the quantizer halves the value of Δ, which reduces the noise power by the factor ¼. In terms of decibels, the noise power change is $10 \cdot \log_{10}\left(\frac{1}{4}\right) \approx -6 \text{ dB}.10$.

Because the set of possible output values of a quantizer is countable, any quantizer can be decomposed into two distinct stages, which can be referred to as the *classification* stage (or *forward quantization* stage) and the *reconstruction* stage (or *inverse quantization* stage), where the classification stage maps the input value to an integer *quantization index* k and the reconstruction stage maps the index k to the *reconstruction value* y_k that is the output approximation of the input value. For the example uniform quantizer described above, the forward quantization stage can be expressed as,

$$k = \left\lfloor \frac{x}{\Delta} + \frac{1}{2} \right\rfloor,$$

and the reconstruction stage for this example quantizer is simply,

$$y_k = k \cdot \Delta.$$

This decomposition is useful for the design and analysis of quantization behavior, and it illustrates how the quantized data can be communicated over a communication channel – a *source encoder* can perform the forward quantization stage and send the index information through a communication channel, and a *decoder* can perform the reconstruction stage to produce the output approximation of the original input data. In general, the forward quantization stage may use any function that maps the input data to the integer space of the quantization index data, and the inverse quantization stage can conceptually (or literally) be a table look-up operation to map each quantization index to a corresponding reconstruction value. This two-stage decomposition applies equally well to vector as well as scalar quantizers.

Mid-riser and Mid-tread Uniform Quantizers

Most uniform quantizers for signed input data can be classified as being of one of two types: mid-riser and mid-tread. The terminology is based on what happens in the region around the value 0, and uses the analogy of viewing the input-output function of the quantizer as a stairway. Mid-tread quantizers have a zero-valued reconstruction level (corresponding to a tread of a stairway), while mid-riser quantizers have a zero-valued classification threshold (corresponding to a riser of a stairway).

Mid-tread quantization involves rounding. The formulas for mid-tread uniform quantization are provided in the previous section.

Mid-riser quantization involves truncation. The input-output formula for a mid-riser uniform quantizer is given by,

$$Q(x) = \Delta \cdot \left(\left\lfloor \frac{x}{\Delta} \right\rfloor + \frac{1}{2} \right),$$

where the classification rule is given by,

$$k = \left\lfloor \frac{x}{\Delta} \right\rfloor$$

and the reconstruction rule is,

$$y_k = \Delta \cdot \left(k + \tfrac{1}{2} \right).$$

Note that mid-riser uniform quantizers do not have a zero output value – their minimum output magnitude is half the step size. In contrast, mid-tread quantizers do have a zero output level. For some applications, having a zero output signal representation may be a necessity.

In general, a mid-riser or mid-tread quantizer may not actually be a uniform quantizer – i.e., the size of the quantizer's classification intervals may not all be the same, or the spacing between its possible output values may not all be the same. The distinguishing characteristic of a mid-riser quantizer is that it has a classification threshold value that is exactly zero, and the distinguishing characteristic of a mid-tread quantizer is that is it has a reconstruction value that is exactly zero.

Dead-zone Quantizers

A dead-zone quantizer is a type of mid-tread quantizer with symmetric behavior around 0. The region around the zero output value of such a quantizer is referred to as the *dead zone* or *deadband*. The dead zone can sometimes serve the same purpose as a noise gate or squelch function. Especially for compression applications, the dead-zone may be given a different width than that for the other steps. For an otherwise-uniform quantizer, the dead-zone width can be set to any value by using the forward quantization rule,

$$k = \text{sgn}\,(x)\,\cdot\,\max\left(0, \left\lfloor \frac{|x|-w/2}{\Delta}+1 \right\rfloor\right),$$

where the function ( ) is the sign function (also known as the *signum* function). The general reconstruction rule for such a dead-zone quantizer is given by,

$$y_k = \text{sgn}\,(k)\cdot\left(\frac{w}{2}+\Delta\cdot(\,|k|-1+r_k\,)\right),$$

where r_k is a reconstruction offset value in the range of 0 to 1 as a fraction of the step size. Ordinarily, $0 \le r_k \le \frac{1}{2}$ when quantizing input data with a typical pdf that is symmetric around zero and reaches its peak value at zero (such as a Gaussian, Laplacian, or generalized Gaussian pdf). Although r_k may depend on in general, and can be chosen to fulfill the optimality condition described below, it is often simply set to a constant, such as $\frac{1}{2}$. (Note that in this definition, $y_0 = 0$ due to the definition of the sgn ( ) function, so r_0 has no effect).

A very commonly used special case (e.g., the scheme typically used in financial accounting and elementary mathematics) is to set $w = \Delta$ and $r_k = \frac{1}{2}$ for all k In this case, the dead-zone quantizer is also a uniform quantizer, since the central dead-zone of this quantizer has the same width as all of its other steps, and all of its reconstruction values are equally spaced as well.

Granular Distortion and Overload Distortion

Often the design of a quantizer involves supporting only a limited range of possible output values and performing clipping to limit the output to this range whenever the input exceeds the supported range. The error introduced by this clipping is referred to as overload distortion. Within the extreme limits of the supported range, the amount of spacing between the selectable output values of a quantizer is referred to as its granularity, and the error introduced by this spacing is referred to as granular distortion. It is common for the design of a quantizer to involve determining the proper balance between granular distortion and overload distortion. For a given supported number of possible output values, reducing the average granular distortion may involve increasing the average overload distortion, and vice versa. A technique for controlling the amplitude of the signal (or, equivalently, the quantization step size Δ) to achieve the appropriate balance is the use of automatic gain control (AGC). However, in some quantizer designs, the concepts of granular error and overload error may not apply (e.g., for a quantizer with a limited range of input data or with a countably infinite set of selectable output values).

The Additive noise Model for Quantization Error

A common assumption for the analysis of quantization error is that it affects a signal processing system in a similar manner to that of additive white noise – having negligible correlation with the signal and an approximately flat power spectral density. The additive noise model is commonly used for the analysis of quantization error effects in digital filtering systems, and it can be very useful in such analysis. It has been shown to be a valid model in cases of high resolution quantization (small Δ relative to the signal strength) with smooth probability density functions.

Additive noise behavior is not always a valid assumption. Quantization error (for quantizers defined as described here) is deterministically related to the signal and not entirely independent of it. Thus, periodic signals can create periodic quantization noise. And in some cases it can even cause limit cycles to appear in digital signal processing systems. One way to ensure effective independence of the quantization error from the source signal is to perform dithered quantization (sometimes with noise shaping), which involves adding random (or pseudo-random) noise to the signal prior to quantization.

Quantization Error Models

In the typical case, the original signal is much larger than one least significant bit (LSB). When this is the case, the quantization error is not significantly correlated with the signal, and has an approximately uniform distribution. In the rounding case, the quantization error has a mean of zero and the RMS value is the standard deviation of this distribution, given by $\frac{1}{\sqrt{12}}\text{LSB} \approx 0.289\text{LSB}$.

In the truncation case the error has a non-zero mean of $\frac{1}{2}\text{LSB}$ and the RMS value is $\frac{1}{\sqrt{3}}\text{LSB}$.

In either case, the standard deviation, as a percentage of the full signal range, changes by a factor of 2 for each 1-bit change in the number of quantizer bits. The potential signal-to-quantization-noise power ratio therefore changes by 4, or $10 \cdot \log_{10}(4) = 6.02$ decibels per bit.

At lower amplitudes the quantization error becomes dependent on the input signal, resulting in distortion. This distortion is created after the anti-aliasing filter, and if these distortions are above 1/2 the sample rate they will alias back into the band of interest. In order to make the quantization error independent of the input signal, noise with an amplitude of 2 least significant bits is added to the signal. This slightly reduces signal to noise ratio, but, ideally, completely eliminates the distortion. It is known as dither.

Quantization Noise Model

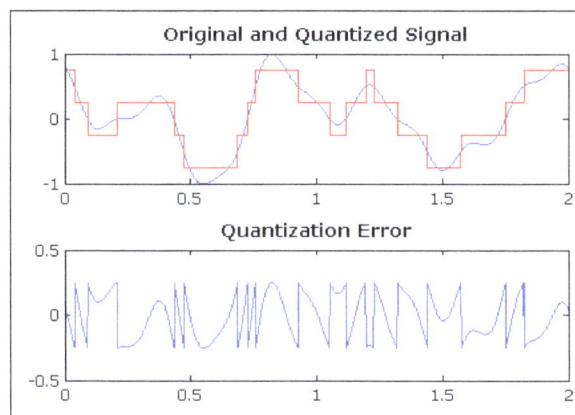

Quantization noise for a 2-bit ADC operating at infinite sample rate. The difference between the blue and red signals in the upper graph is the quantization error, which is "added" to the quantized signal and is the source of noise.

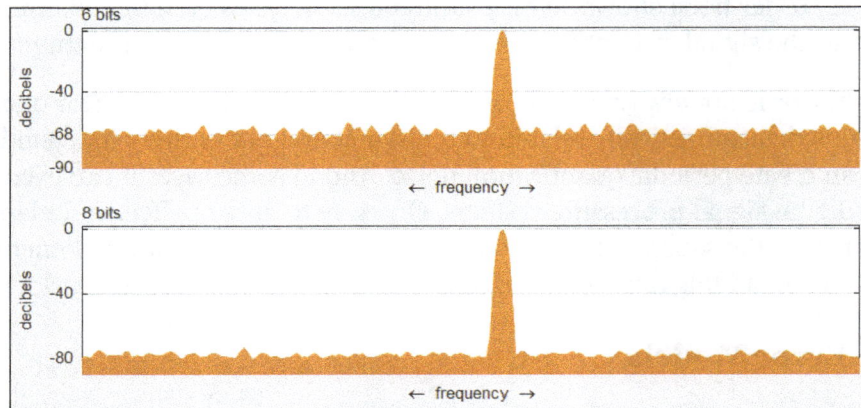

Comparison of quantizing a sinusoid to 64 levels (6 bits) and 256 levels (8 bits). The additive noise created by 6-bit quantization is 12 dB greater than the noise created by 8-bit quantization.

When the spectral distribution is flat, as in this example, the 12 dB difference manifests as a measurable difference in the noise floors.

Quantization noise is a model of quantization error introduced by quantization in the analog-to-digital conversion (ADC) in telecommunication systems and signal processing. It is a rounding error between the analog input voltage to the ADC and the output digitized value. The noise is non-linear and signal-dependent. It can be modelled in several different ways.

In an ideal analog-to-digital converter, where the quantization error is uniformly distributed between −1/2 LSB and +1/2 LSB, and the signal has a uniform distribution covering all quantization levels, the Signal-to-quantization-noise ratio (SQNR) can be calculated from,

$$\text{SQNR} = 20\log_{10}(2^Q) \approx 6.02.Q \text{ dB}$$

Where Q is the number of quantization bits.

The most common test signals that fulfill this are full amplitude triangle waves and sawtooth waves.

For example, a 16-bit ADC has a maximum signal-to-noise ratio of 6.02 × 16 = 96.3 dB.

When the input signal is a full-amplitude sine wave the distribution of the signal is no longer uniform, and the corresponding equation is instead,

$$\text{SQNR} \approx 1.761 + 6.02.Q \text{ dB}$$

Here, the quantization noise is once again *assumed* to be uniformly distributed. When the input signal has a high amplitude and a wide frequency spectrum this is the case. In this case a 16-bit ADC has a maximum signal-to-noise ratio of 98.09 dB. The 1.761 difference in signal-to-noise only occurs due to the signal being a full-scale sine wave instead of a triangle/sawtooth.

Quantization noise power can be derived from,

$$N = \frac{(\delta v)^2}{12} W$$

where δv is the voltage of the level.

Typical real-life values are worse than this theoretical minimum, due to the addition of dither to reduce the objectionable effects of quantization, and to imperfections of the ADC circuitry.

For complex signals in high-resolution ADCs this is an accurate model. For low-resolution ADCs, low-level signals in high-resolution ADCs, and for simple waveforms the quantization noise is not uniformly distributed, making this model inaccurate. In these cases the quantization noise distribution is strongly affected by the exact amplitude of the signal.

The calculations above, however, assume a completely filled input channel. If this is not the case - if the input signal is small - the relative quantization distortion can be very large. To circumvent this issue, analog compressors and expanders can be used, but these introduce large amounts of distortion as well, especially if the compressor does not match the expander. The application of such compressors and expanders is also known as companding.

Rate–distortion Quantizer Design

A scalar quantizer, which performs a quantization operation, can ordinarily be decomposed into two stages:

- Classification: A process that classifies the input signal range into M non-overlapping intervals $\{I_k\}_{k=1}^{M}$, by defining M- 1 boundary (decision) values $\{b_k\}_{k=1}^{M-1}$, such that $I_k = [b_{k-1}, b_k)$ for $k = 1, 2, ..., M$, with the extreme limits defined by $b_0 = -\infty$ and $b_M = \infty$. All the inputs x that fall in a given interval range I_k are associated with the same quantization index k.

- Reconstruction: Each interval I_k is represented by a reconstruction value y_k which implements the mapping $x \in I_k \Rightarrow y = y_k$.

These two stages together comprise the mathematical operation of $y = Q(x)$.

Entropy coding techniques can be applied to communicate the quantization indices from a source encoder that performs the classification stage to a decoder that performs the reconstruction stage. One way to do this is to associate each quantization index k with a binary codeword c_k. An important consideration is the number of bits used for each codeword denoted here by length (c_k).

As a result, the design of an M- level quantizer and an associated set of codewords for communicating its index values requires finding the values of, $\{b_k\}_{k=1}^{M-1}, \{c_k\}_{k=1}^{M}$ and $\{y_k\}_{k=1}^{M}$ which optimally satisfy a selected set of design constraints such as the bit rate R and distortion D.

Assuming that an information source S produces random variables X with an associated probability density function $f(x)$, the probability p_k that the random variable falls within a particular quantization interval I_k is given by,

$$p_k = P[x \in I_k] = \int_{b_{k-1}}^{b_k} f(x)dx.$$

The resulting bit rate R , in units of average bits per quantized value, for this quantizer can be derived as follows:

$$R=\sum_{k=1}^{M} p_k \times length(c_k) = \sum_{k=1}^{M} length(c_k) \int_{b_{k-1}}^{b_k} f(x) dx \ .$$

If it is assumed that distortion is measured by mean squared error, the distortion D, is given by:

$$D = E[\ (x-Q(x)\ )^2\ ] = \int_{\infty}^{\infty} (\ x-Q(x)\ )^2\ f(x)\ dx = \sum_{k=1}^{M} \int_{b_{k-1}}^{b_k} (x-y_k)^2 f(x) dx$$

Note that other distortion measures can also be considered, although mean squared error is a popular one.

A key observation is that rate R depends on the decision boundaries $\{b_k\}_{k=1}^{M-1}$ and the codeword lengths $\{length\ (c_k)\}_{k=1}^{M}$, whereas the distortion D depends on the decision boundaries $\{b_k\}_{k=1}^{M-1}$ and the reconstruction levels $\{y_k\}_{k=1}^{M}$.

After defining these two performance metrics for the quantizer, a typical Rate–Distortion formulation for a quantizer design problem can be expressed in one of two ways:

1.   Given a maximum distortion constraint $D \leq D_{max}$, minimize the bit rate R .

2.   Given a maximum bit rate constraint $R \leq R_{max}$, minimize the distortion D .

Often the solution to these problems can be equivalently (or approximately) expressed and solved by converting the formulation to the unconstrained problem $\min\{D+\lambda\times R\}$ where the Lagrange multiplier λ is a non-negative constant that establishes the appropriate balance between rate and distortion. Solving the unconstrained problem is equivalent to finding a point on the convex hull of the family of solutions to an equivalent constrained formulation of the problem. However, finding a solution – especially a closed-form solution – to any of these three problem formulations can be difficult. Solutions that do not require multi-dimensional iterative optimization techniques have been published for only three probability distribution functions: the uniform, exponential, and Laplacian distributions. Iterative optimization approaches can be used to find solutions in other cases.

Note that the reconstruction values $\{\ y_k\ \}_{k=1}^{M}$ affect only the distortion – they do not affect the bit rate – and that each individual y_k makes a separate contribution d_k to the total distortion as shown below,

$$D = \sum_{k=1}^{M} d_k$$

where,

$$d_k = \int_{b_{k-1}}^{b_k} (x - y_k)^2\ f(x)\ dx$$

This observation can be used to ease the analysis – given the set of $\{b_k\}_{k=1}^{M-1}$ values, the value of each y_k can be optimized separately to minimize its contribution to the distortion D.

For the mean-square error distortion criterion, it can be easily shown that the optimal set of reconstruction values $\{y_k^*\}_{k=1}^M$ is given by setting the reconstruction value y_k within each interval I_k to the conditional expected value (also referred to as the *centroid*) within the interval, as given by:

$$y_k^* = \frac{1}{p_k}\int_{b_{k-1}}^{b_k} xf(x)\,dx.$$

The use of sufficiently well-designed entropy coding techniques can result in the use of a bit rate that is close to the true information content of the indices $\{k\}_{k=1}^M$, such that effectively,

$$\text{length}(c_k) \approx -\log_2(p_k)$$

and therefore,

$$R = \sum_{k=1}^M -p_k \cdot \log_2(p_k).$$

The use of this approximation can allow the entropy coding design problem to be separated from the design of the quantizer itself. Modern entropy coding techniques such as arithmetic coding can achieve bit rates that are very close to the true entropy of a source, given a set of known (or adaptively estimated) probabilities $\{p_k\}_{k=1}^M$.

In some designs, rather than optimizing for a particular number of classification regions M, the quantizer design problem may include optimization of the value of M as well. For some probabilistic source models, the best performance may be achieved when M approaches infinity.

Neglecting the Entropy Constraint: Lloyd–Max Quantization

In the above formulation, if the bit rate constraint is neglected by setting λ equal to 0, or equivalently if it is assumed that a fixed-length code (FLC) will be used to represent the quantized data instead of a variable-length code (or some other entropy coding technology such as arithmetic coding that is better than an FLC in the rate–distortion sense), the optimization problem reduces to minimization of distortion D alone.

The indices produced by an M-level quantizer can be coded using a fixed-length code using $R = \lceil \log_2 M \rceil$ bits/symbol. For example, when $M = 256$ levels, the FLC bit rate R is 8 bits/symbol. For this reason, such a quantizer has sometimes been called an 8-bit quantizer. However using an FLC eliminates the compression improvement that can be obtained by use of better entropy coding.

Assuming an FLC with M levels, the Rate–Distortion minimization problem can be reduced to distortion minimization alone. The reduced problem can be stated as follows: given a source X with pdf $f(x)$ and the constraint that the quantizer must use only M classification regions, find the decision boundaries $\{b_k\}_{k=1}^{M-1}$ and reconstruction levels $\{y_k\}_{k=1}^M$ to minimize the resulting distortion,

$$D = E[(x-Q(x))^2] = \int_{-\infty}^{\infty}(x-Q(x))^2 f(x)\,dx = \sum_{k=1}^M \int_{b_{k-1}}^{b_k}(x-y_k)^2 f(x)\,dx = \sum_{k=1}^M d_k.$$

Finding an optimal solution to the above problem results in a quantizer sometimes called a MMSQE (minimum mean-square quantization error) solution, and the resulting pdf-optimized (non-uniform) quantizer is referred to as a *Lloyd–Max* quantizer, named after two people who independently developed iterative methods to solve the two sets of simultaneous equations resulting from $\partial D/\partial b_k = 0$ and $\partial D/\partial y_k = 0$ as follows,

$$\frac{\partial D}{\partial b_k} = 0 \Rightarrow b_k = \frac{y_k + y_{k+1}}{2},$$

which places each threshold at the midpoint between each pair of reconstruction values, and

$$\frac{\partial D}{\partial y_k} = 0 \Rightarrow y_k = \frac{\int_{b_{k-1}}^{b_k} xf(x)\,dx}{\int_{b_{k-1}}^{b_k} f(x)\,dx} = \frac{1}{p_k}\int_{b_{k-1}}^{b_k} xf(x)\,dx$$

which places each reconstruction value at the centroid (conditional expected value) of its associated classification interval.

Lloyd's Method I algorithm, originally described in 1957, can be generalized in a straightforward way for application to vector data. This generalization results in the Linde–Buzo–Gray (LBG) or k-means classifier optimization methods. Moreover, the technique can be further generalized in a straightforward way to also include an entropy constraint for vector data.

Uniform Quantization and the 6 dB/bit Approximation

The Lloyd–Max quantizer is actually a uniform quantizer when the input pdf is uniformly distributed over the range $[y_1 - \Delta/2, y_M + \Delta/2]$. However, for a source that does not have a uniform distribution, the minimum-distortion quantizer may not be a uniform quantizer.

The analysis of a uniform quantizer applied to a uniformly distributed source can be summarized in what follows:

A symmetric source X can be modelled with $f(x) = \frac{1}{2X_{max}}$, for $x \in [-X_{max}, X_{max}]$, and 0 elsewhere. The step size $\Delta = \frac{2X_{max}}{M}$ and the *signal to quantization noise ratio* (SQNR) of the quantizer is,

$$SQNR = 10\log_{10}\frac{\sigma_x^2}{\sigma_q^2} = 10\log_{10}\frac{(M\Delta)^2/12}{\Delta^2/12} = 10\log_{10}M^2 = 20\log_{10}M.$$

For a fixed-length code using N bits, $M = 2^N$,

resulting in $SQNR = 20\log_{10}2^N = N\cdot(20\log_{10}2) = N\cdot6.0206\,dB$, or approximately 6 dB per bit. For example, for N = 8 bits, M = 256 levels and SQNR = 8×6 = 48 dB; and for N = 16 bits, M = 65536 and SQNR = 16×6 = 96 dB. The property of 6 dB improvement in SQNR for each extra bit used in quantization is a well-known figure of merit. However, it must be used with care: this derivation is only for a uniform quantizer applied to a uniform source.

For other source pdfs and other quantizer designs, the SQNR may be somewhat different from that predicted by 6 dB/bit, depending on the type of pdf, the type of source, the type of quantizer, and the bit rate range of operation.

However, it is common to assume that for many sources, the slope of a quantizer SQNR function can be approximated as 6 dB/bit when operating at a sufficiently high bit rate. At asymptotically high bit rates, cutting the step size in half increases the bit rate by approximately 1 bit per sample (because 1 bit is needed to indicate whether the value is in the left or right half of the prior double-sized interval) and reduces the mean squared error by a factor of 4 (i.e., 6 dB) based on the $\Delta^2/12$ approximation.

At asymptotically high bit rates, the 6 dB/bit approximation is supported for many source pdfs by rigorous theoretical analysis. Moreover, the structure of the optimal scalar quantizer (in the rate–distortion sense) approaches that of a uniform quantizer under these conditions.

For the samples that are highly correlated, when encoded by PCM technique, leave redundant information behind. To process this redundant information and to have a better output, it is a wise decision to take a predicted sampled value, assumed from its previous output and summarize them with the quantized values. Such a process is called as Differential PCM (DPCM) technique.

DPCM Transmitter

The DPCM Transmitter consists of Quantizer and Predictor with two summer circuits. Following is the block diagram of DPCM transmitter.

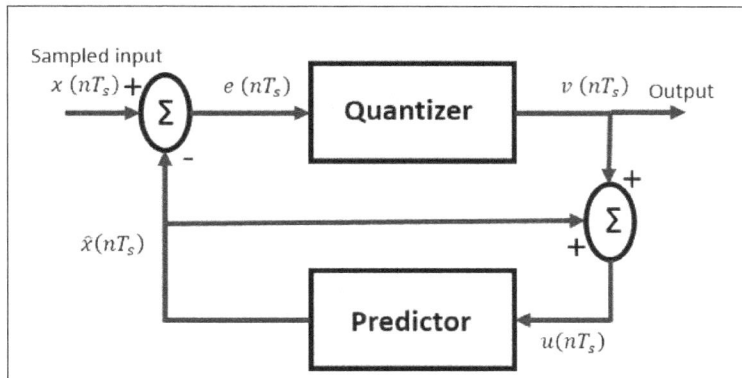

The signals at each point are named as:

* $x(nT_s)$ is the sampled input.

* $\hat{x}(nT_s)$ is the predicted sample.

* $e(nT_s)$ is the difference of sample d input and predicted output, often called as prediction error.

* $v(nT_s)$ is the quantized output.

* $u(nT_s)$ is the predictor input which is actually the summer output of the predictor output and the quantizer output.

The predictor produces the assumed samples from the previous outputs of the transmitter circuit. The input to this predictor is the quantized versions of the input signal $x(nT_s)$.

Quantizer Output is represented as:

$$v(nT_s) = Q[e(nT_s)]$$

$$= e(nT_s) + q(nT_s)$$

Where $q(nT_s)$ is the quantization error.

Predictor input is the sum of quantizer output and predictor output,

$$u(nT_s) = \hat{x}(nT_s) + v(nT_s)$$

$$u(nT_s) = \hat{x}(nT_s) + e(nT_s) + q(nT_s)$$

$$u(nT_s) = x(nT_s) + q(nT_s)$$

The same predictor circuit is used in the decoder to reconstruct the original input.

DPCM Receiver

The block diagram of DPCM Receiver consists of a decoder, a predictor, and a summer circuit. Following is the diagram of DPCM Receiver.

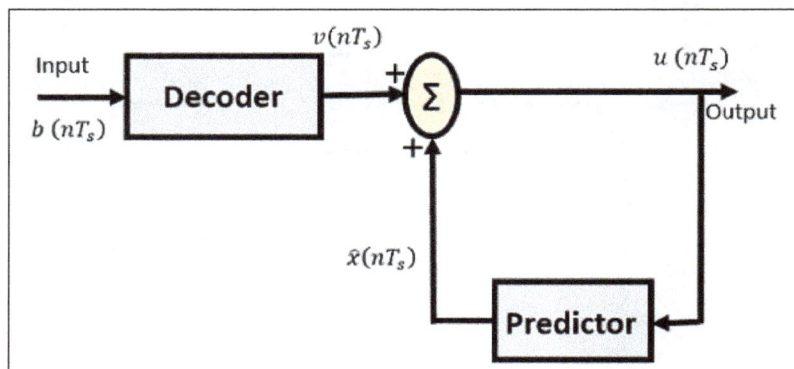

The notation of the signals is the same as the previous ones. In the absence of noise, the encoded receiver input will be the same as the encoded transmitter output. the predictor assumes a value, based on the previous outputs. The input given to the decoder is processed and that output is summed up with the output of the predictor, to obtain a better output.

The sampling rate of a signal should be higher than the Nyquist rate, to achieve better sampling. If this sampling interval in Differential PCM is reduced considerably, the sampleto-sample amplitude difference is very small, as if the difference is 1-bit quantization, then the step-size will be very small i.e Δ (delta).

Delta Modulation

The type of modulation, where the sampling rate is much higher and in which the stepsize after quantization is of a smaller value Δ, such a modulation is termed as delta modulation.

Features of Delta Modulation

Following are some of the features of delta modulation:

- An over-sampled input is taken to make full use of the signal correlation.

- The quantization design is simple.

- The input sequence is much higher than the Nyquist rate.

- The quality is moderate.

- The design of the modulator and the demodulator is simple.

- The stair-case approximation of output waveform.

- The step-size is very small, i.e., Δ (delta).

- The bit rate can be decided by the user.

- This involves simpler implementation.

Delta Modulation is a simplified form of DPCM technique, also viewed as 1-bit DPCM scheme. As the sampling interval is reduced, the signal correlation will be higher.

Delta Modulator

The Delta Modulator comprises of a 1-bit quantizer and a delay circuit along with two summer circuits. Following is the block diagram of a delta modulator.

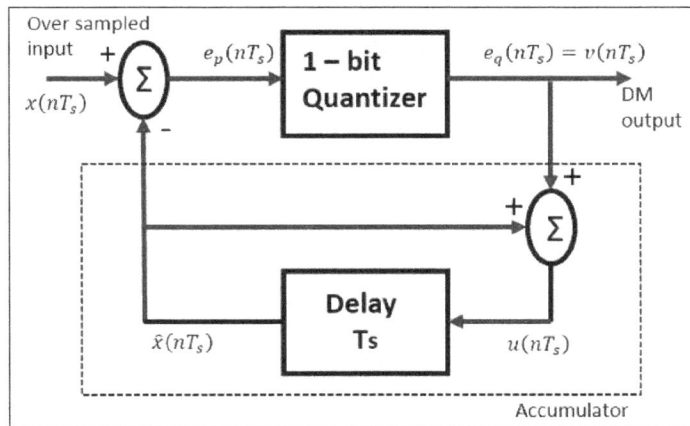

The predictor circuit in DPCM is replaced by a simple delay circuit in DM. From the above diagram, we have the notations as:

- $x(nT_s)$ = over sampled input.

- $e_p(nT_s)$ = summer output and quantizer input.

- $e_q(nT_s)$ = quantizer output = v(nTs).

- $\hat{x}(nT_s)$ = output of delay circuit.

- $u(nT_s)$ = input of delay circuit.

Using these notations, now we shall try to figure out the process of delta modulation.

$$e_p(nT_s) \;=\; x\,(nT_s)\; -\; \hat{x}\,(nT_s)$$

$$= x\,(nT_s)\; -u\,(\,[n\text{-}1]\;T_s)$$

$$= x(nT_s) - [\hat{x}[[n-1]T_s] + v[[n-1]T_s]]$$

Further,

$$v(nT_s) = e_q(nT_s) = S.sig.[e_p(nT_s)]$$

$$u(nT_s) = \hat{x}(nT_s) + e_q(nT_s)$$

where,

$\hat{x}(nT_s)$ = the previous value of the delay circuit.

$e_q(nT_s)$ = quantizer output = $v(nT_s)$.

Hence,

$$u(nT_s) = u([n-1]T_s) + v(nT_s)$$

Which means,

The present input of the delay unit = (The previous output of the delay unit) + (the present quantizer output)

Assuming zero condition of accumulation,

$$u(nT_s) = S\sum_{j=1}^{n} sig[e_p(jT_s)]$$

Accumulated version of DM output $\displaystyle\sum_{j=1}^{n}v(jT_s)$

Now, note that,

$$\hat{x}(nT_s) = u([n-1]T_s)$$

$$= \sum_{j=1}^{n-1} v(jTs)$$

Delay unit output is an Accumulator output lagging by one sample.

From equations $\sum_{j=1}^{n} v(jT_s)$ & $= \sum_{j=1}^{n-1} v(jTs)$, we get a possible structure for the demodulator.

A Stair-case approximated waveform will be the output of the delta modulator with the step-size as delta (Δ). The output quality of the waveform is moderate.

Delta Demodulator

The delta demodulator comprises of a low pass filter, a summer, and a delay circuit. The predictor circuit is eliminated here and hence no assumed input is given to the demodulator.

Following is the diagram for delta demodulator.

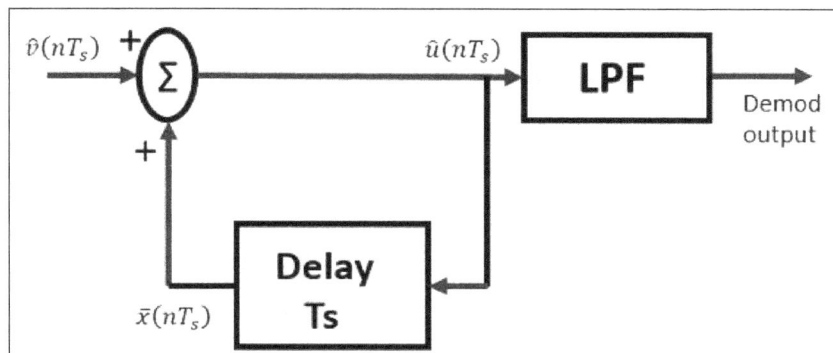

From the above diagram, we have the notations as:

- $\hat{v}(nT_s)$ is the input sample

- $\hat{u}(nT_s)$ is the summer output

- $\bar{x}(nT_s)$ is the delayed output

A binary sequence will be given as an input to the demodulator. The stair-case approximated output is given to the LPF.

Low pass filter is used for many reasons, but the prominent reason is noise elimination for out-of-band signals. The step-size error that may occur at the transmitter is called granular noise, which is eliminated here. If there is no noise present, then the modulator output equals the demodulator input.

Advantages of DM over DPCM

- 1-bit quantizer.

- Very easy design of the modulator and the demodulator.

However, there exists some noise in DM.

- Slope Over load distortion (when Δ is small).

- Granular noise (when Δ is large).

Adaptive Delta Modulation

Following is the block diagram of Adaptive delta modulator.

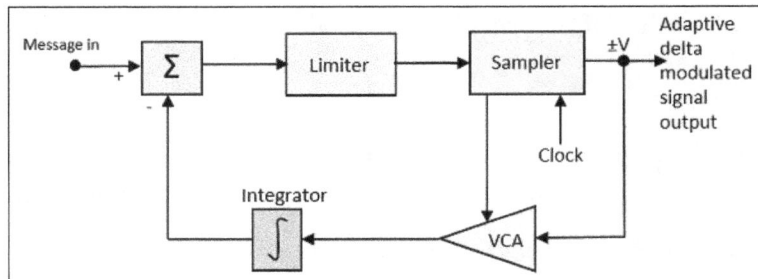

Adaptive delta modulator.

The gain of the voltage controlled amplifier is adjusted by the output signal from the sampler. The amplifier gain determines the step-size and both are proportional.

ADM quantizes the difference between the value of the current sample and the predicted value of the next sample. It uses a variable step height to predict the next values, for the faithful reproduction of the fast varying values.

Techniques

There are a few techniques which have paved the basic path to digital communication processes. For the signals to get digitized, we have the sampling and quantizing techniques.

For them to be represented mathematically, we have LPC and digital multiplexing techniques. These digital modulation techniques are further discussed.

Linear Predictive Coding

Linear Predictive Coding (LPC) is a tool which represents digital speech signals in linear predictive model. This is mostly used in audio signal processing, speech synthesis, speech recognition, etc.

Linear prediction is based on the idea that the current sample is based on the linear combination of past samples. The analysis estimates the values of a discrete-time signal as a linear function of the previous samples.

The spectral envelope is represented in a compressed form, using the information of the linear predictive model. This can be mathematically represented as:

$$s(n) = \sum_{k=1}^{p} a_k s(n-k) \text{ for some value of p and } \alpha_k.$$

Where,

- s(n) is the current speech sample.

- k is a particular sample.

- p is the most recent value.

- α_k is the predictor co-efficient.
- $s(n - k)$ is the previous speech sample.

For LPC, the predictor co-efficient values are determined by minimizing the sum of squared differences (over a finite interval) between the actual speech samples and the linearly predicted ones.

This is a very useful method for encoding speech at a low bit rate. The LPC method is very close to the Fast Fourier Transform (FFT) method.

Multiplexing

Multiplexing is the process of combining multiple signals into one signal, over a shared medium. If digital signals are multiplexed, it is called as digital multiplexing.

Multiplexing was first developed in telephony. A number of signals were combined to send through a single cable. The process of multiplexing divides a communication channel into several number of logical channels, allotting each one for a different message signal or a data stream to be transferred. The device that does multiplexing, can be called as a MUX. The reverse process, i.e., extracting the number of channels from one, which is done at the receiver is called as de-multiplexing. The device which does de-multiplexing is called as DEMUX.

The following figures represent MUX and DEMUX. Their primary use is in the field of communications.

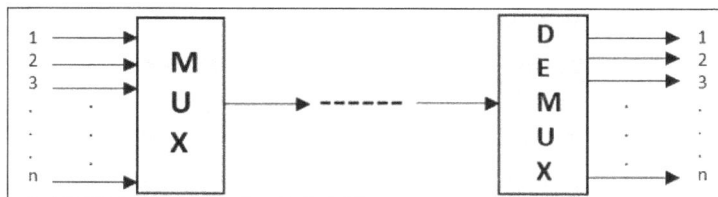

Multiplexing and Demultiplexing.

Types of Multiplexers

There are mainly two types of multiplexers, namely analog and digital. They are further divided into FDM, WDM, and TDM. The following figure gives a detailed idea on this classification.

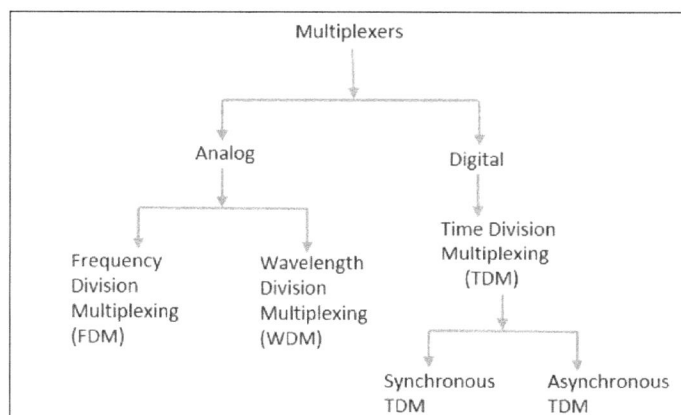

Actually, there are many types of multiplexing techniques. Of them all, we have the main types with general classification, mentioned in the above figure.

Analog Multiplexing

The analog multiplexing techniques involve signals which are analog in nature. The analog signals are multiplexed according to their frequency (FDM) or wavelength (WDM).

Frequency Division Multiplexing

In analog multiplexing, the most used technique is Frequency Division Multiplexing (FDM). This technique uses various frequencies to combine streams of data, for sending them on a communication medium, as a single signal.

Example – A traditional television transmitter, which sends a number of channels through a single cable, uses FDM.

Wavelength Division Multiplexing

Wavelength Division multiplexing is an analog technique, in which many data streams of different wavelengths are transmitted in the light spectrum. If the wavelength increases, the frequency of the signal decreases. A prism which can turn different wavelengths into a single line, can be used at the output of MUX and input of DEMUX.

Example – Optical fiber communications use WDM technique to merge different wavelengths into a single light for communication.

Line Code

A line code is the code used for data transmission of a digital signal over a transmission line. This process of coding is chosen so as to avoid overlap and distortion of signal such as inter-symbol interference.

Properties of Line Coding

Following are the properties of line coding:

- As the coding is done to make more bits transmit on a single signal, the bandwith used is much reduced.
- For a given bandwith, the power is efficiently used.
- The probability of error is much reduced.
- Error detection is done and the bipolar too has a correction capability.
- Power density is much favorable.
- The timing content is adequate.
- Long strings of 1s and 0s is avoided to maintain transparency.

Types of Line Coding

There are 3 types of Line Coding:

- Unipolar,
- Polar,
- Bi-polar.

Unipolar Signaling

Unipolar signaling is also called as On-Off Keying or simply OOK.

The presence of pulse represents a 1 and the absence of pulse represents a 0.

There are two variations in Unipolar signaling:

- Non Return to Zero (NRZ).
- Return to Zero (RZ).

Unipolar Non-Return to Zero

In this type of unipolar signaling, a High in data is represented by a positive pulse called as Mark, which has a duration T_0 equal to the symbol bit duration. A Low in data input has no pulse.

The following figure clearly depicts this.

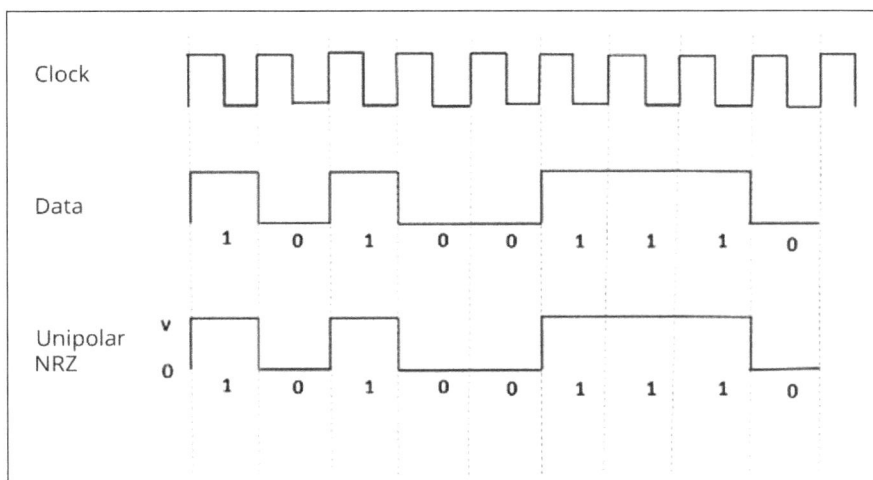

Advantages

The advantages of Unipolar NRZ are:

- It is simple.
- A lesser bandwith is required.

Disadvantages

The disadvantages of Unipolar NRZ are:

- No error correction done.

- Presence of low frequency components may cause the signal droop.

- No clock is present.

- Loss of synchronization is likely to occur (especially for long strings of 1s and 0s).

Unipolar Return to Zero

In this type of unipolar signaling, a High in data, though represented by a Mark pulse, its duration To is less than the symbol bit duration. Half of the bit duration remains high but it immediately returns to zero and shows the absence of pulse during the remaining half of the bit duration.

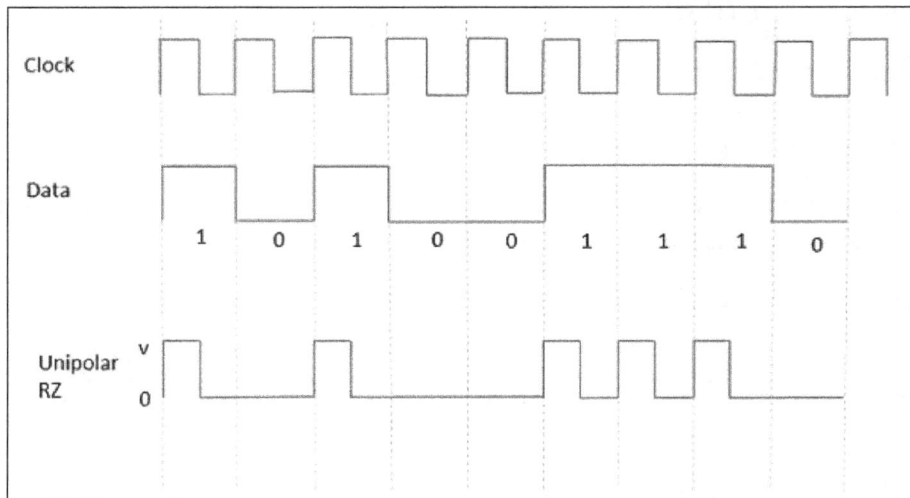

Advantages

The advantages of Unipolar RZ are:

- It is simple.

- The spectral line present at the symbol rate can be used as a clock.

Disadvantages

The disadvantages of Unipolar RZ are:

- No error correction.

- Occupies twice the bandwith as unipolar NRZ.

- The signal droop is caused at the places where signal is non-zero at 0 Hz.

Polar Signaling

There are two methods of Polar Signaling. They are:

- Polar NRZ,
- Polar RZ.

Polar NRZ

In this type of Polar signaling, a High in data is represented by a positive pulse, while a Low in data is represented by a negative pulse. The following figure depicts this well.

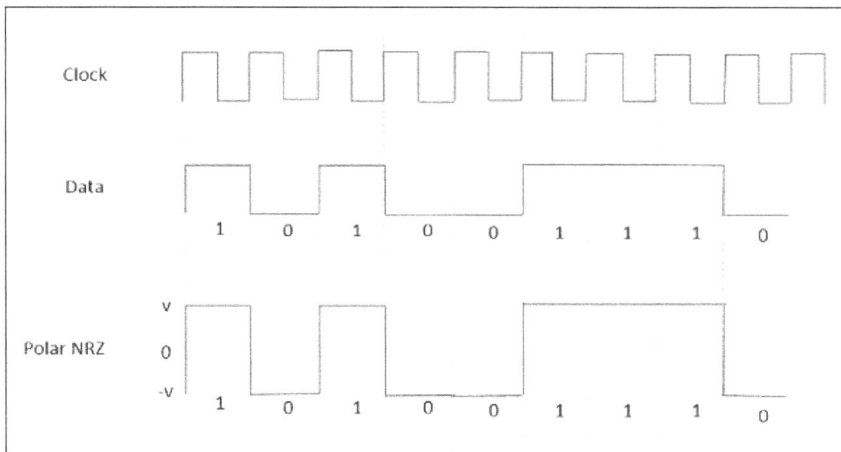

Advantages

The advantages of Polar NRZ are –

- It is simple.
- No low-frequency components are present.

Disadvantages

The disadvantages of Polar NRZ are –

- No error correction.
- No clock is present.
- The signal droop is caused at the places where the signal is non-zero at 0 Hz.

Polar RZ

In this type of Polar signaling, a High in data, though represented by a Mark pulse, its duration T_o is less than the symbol bit duration. Half of the bit duration remains high but it immediately returns to zero and shows the absence of pulse during the remaining half of the bit duration.

However, for a Low input, a negative pulse represents the data, and the zero level remains same for the other half of the bit duration. The following figure depicts this clearly.

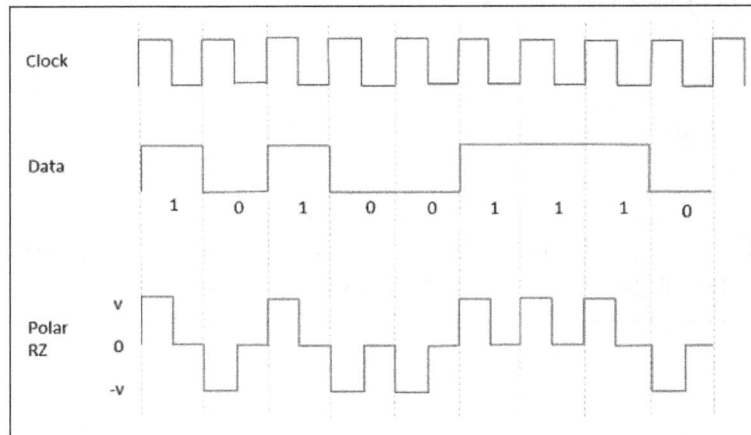

Advantages

The advantages of Polar RZ are:

- It is simple.

- No low-frequency components are present.

Disadvantages

The disadvantages of Polar RZ are:

- No error correction.

- No clock is present.

- Occupies twice the bandwith of Polar NRZ.

- The signal droop is caused at places where the signal is non-zero at 0 Hz.

Bipolar Signaling

This is an encoding technique which has three voltage levels namely +, - and 0. Such a signal is called as duo-binary signal.

An example of this type is Alternate Mark Inversion (AMI). For a 1, the voltage level gets a transition from + to − or from − to +, having alternate 1s to be of equal polarity. A 0 will have a zero voltage level.

Even in this method, we have two types:

- Bipolar NRZ,

- Bipolar RZ,

From the models so far discussed, we have learnt the difference between NRZ and RZ. It just goes in the same way here too. The following figure clearly depicts this.

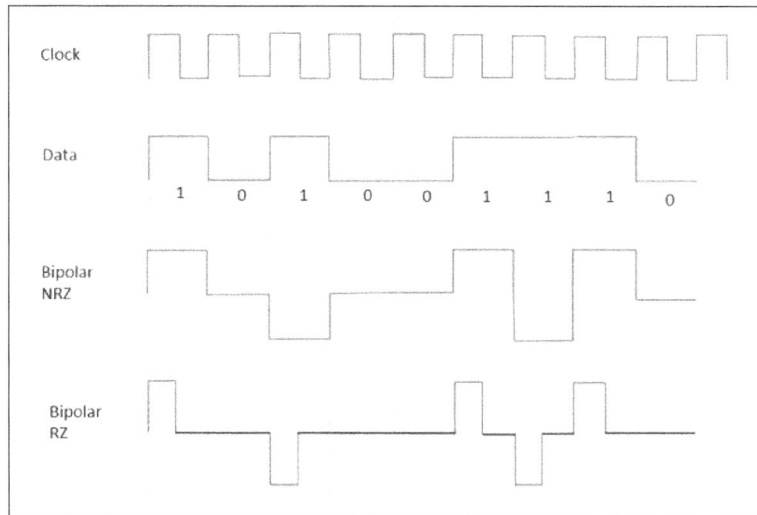

The above figure has both the Bipolar NRZ and RZ waveforms. The pulse duration and symbol bit duration are equal in NRZ type, while the pulse duration is half of the symbol bit duration in RZ type.

Advantages

Following are the advantages:

- It is simple.

- No low-frequency components are present.

- Occupies low bandwith than unipolar and polar NRZ schemes.

- This technique is suitable for transmission over AC coupled lines, as signal drooping doesn't occur here.

- A single error detection capability is present in this.

Disadvantages

Following are the disadvantages:

- No clock is present.

- Long strings of data causes loss of synchronization.

Power Spectral Density

The function which describes how the power of a signal got distributed at various frequencies, in the frequency domain is called as Power Spectral Density (PSD).

PSD is the Fourier Transform of Auto-Correlation (Similarity between observations). It is in the form of a rectangular pulse.

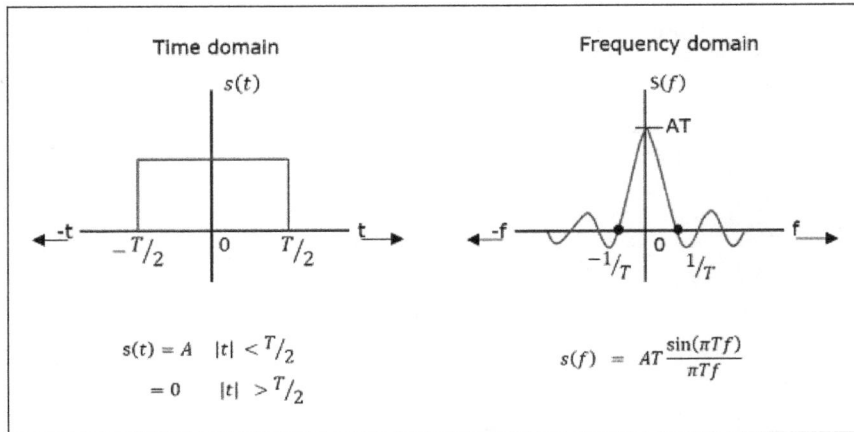

PSD Derivation

According to the Einstein-Wiener-Khintchine theorem, if the auto correlation function or power spectral density of a random process is known, the other can be found exactly.

Hence, to derive the power spectral density, we shall use the time auto-correlation $(R_x(\tau))$ of a power signal $x(t)$ as shown below.

$$R_x(\tau) = lim T_p \to \infty \frac{1}{T_p} \int_{\frac{-Tp}{2}}^{\frac{Tp}{2}} x(t)x(t+\tau)dt$$

Since $x(t)$ consists of impulses, $R_x(\tau)$ can be written as,

$$R_x(\tau) = \frac{1}{T} \sum_{n=-\infty}^{\infty} Rn\delta(\tau - nT)$$

Where $R_n = \lim N \to \infty \frac{1}{N} \sum_k a_k a_{k+n}$

Getting to know that $R_n = R_{-n}$ for real signals, we have,

$$S_x(w) = \frac{1}{T}(R_0 + 2\sum_{n=1}^{\infty} R_n cosnwT)$$

Since the pulse filter has the spectrum of $(w) \leftrightarrow f(t)$ we have,

$$s_y(w) = \left| F(w) \right|^2 S_x(w)$$

$$= \frac{\left| F(w) \right|^2}{T}(\sum_{n=-\infty}^{\infty} R_n e^{-jnwTb})$$

$$= \frac{\left| F(w) \right|^2}{T} (R_0 + 2\sum_{n=1}^{\infty} R_n cosnwT)$$

Hence, we get the equation for Power Spectral Density. Using this, we can find the PSD of various line codes.

DIGITAL MODULATION

Modulation is one type of process used to change the properties of a periodic signal which is known as carrier signal with a modulating signal that normally contains data to be transferred. This technique is frequently used in electronics as well as telecommunications. In the 20th century, most of the radio systems used AM or FM for broadcasting. Even though, this method is called as DM (Digital Modulation). Generally, it involves AM (or) FM (or) PM. Therefore digital modulation must be observed the three modulation techniques how they communicating the information in the form of bits. In the rule, the same types of modulator otherwise demodulator can be used as these modulations are used to transmit analog pattern.

What is Digital Modulation?

The term DM stands for digital modulation, and it is a common term for the techniques of modulation. This modulation uses discrete signals for modulating a carrier wave. Indifference, both the amplitude modulation and frequency modulation techniques are analog. Digital modulation removes communication noise as well as provides enhanced strength for the signal intrusion. But, it is not rare to digital modulation schemes for introducing time delay because of the required process. To avoid this, a comfort SST (Secure Stream Technology) audio is designed.

Digital Modulation (DM) gives more capacity of data, high information security, and accessibility of a faster system by enormous quality communication. Therefore, DM techniques have a huge demand due to their capacity for communicating superior amounts of information than AM (analog modulation) techniques.

Types of Digital Modulation

There are several kinds of digital modulation techniques are available based on the requirement which includes the following.

- ASK or Amplitude shift Key,
- FSK or Frequency shift key,
- PSK or Phase shift key.

ASK (Amplitude Shift Keying)

In amplitude shift keying, once the instant amplitude of the carrier signal is changed in quantity toward m(t) message signal. For instance, if we have the modulated carrier (m(t) coswct) then the carrier signal will be coswct. Because the data is an ON/OFF signal, and the output is also an ON/OFF signal wherever the carrier is there when data is 1, as well as the carrier, is not present when data is 0. Therefore this modulation scheme is called as OOK or on/off keying (OOK) otherwise amplitude shift keying or ASK. The applications of ASK mainly include IR remote controls and fiber optic transmitter & receiver.

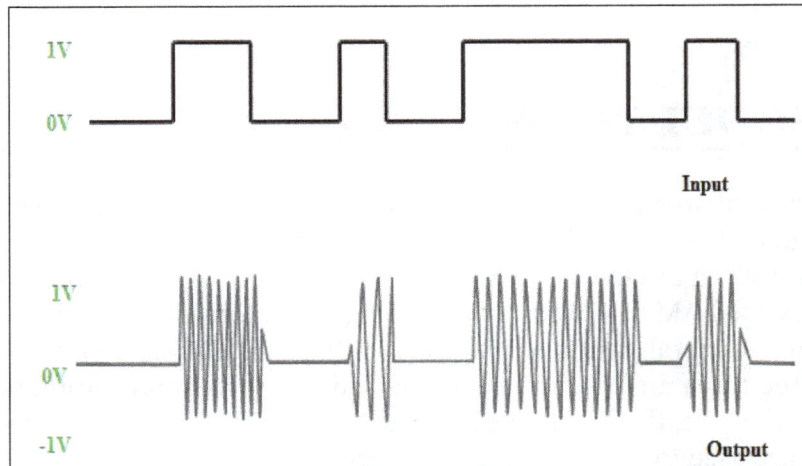

Amplitude Shift Keying.

Frequency Shift Keying

In frequency shift keying, when the immediate frequency of the carrier signal is changed then the information will be transmitted. In this type of modulation, carrier signal has two pre-defined frequencies namely wc1, and wc2. Whenever the data bit is '1' then the carrier signal by wc1 is transmitted that is coswc1. Similarly, when the data bit is '0' then the carrier signal by wc0 will be transmitted that is coswc0. The applications of frequency shift keying mainly include several modems in telemetry systems and phase shift keying.

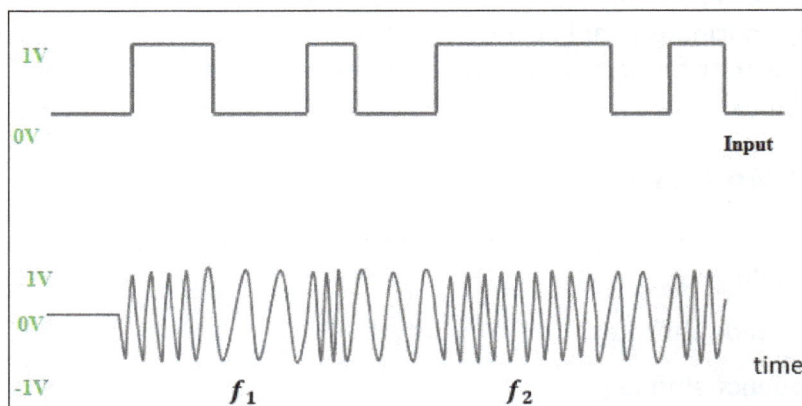

Frequency Shift Keying.

PSK Digital Modulation

In phase shift keying, the instant phase of the carrier signal is moved for this modulation. If the m(t) baseband signal is =1 then the carrier signal within phase will be transmitted. Similarly, If the baseband signal m(t)=0 then the carrier signal by out of phase is transmitted that is cos(wct+Π). If phase shift can be done in four dissimilar quadrants then 2-bit of data will be transmitted at once. This method is an individual case of phase shift keying modulation which is known as Quadrature Phase Shift Keying or QPSK. The applications of phase shift keying include a broadband modem (ADSL), satellite communications, mobile phones, etc.

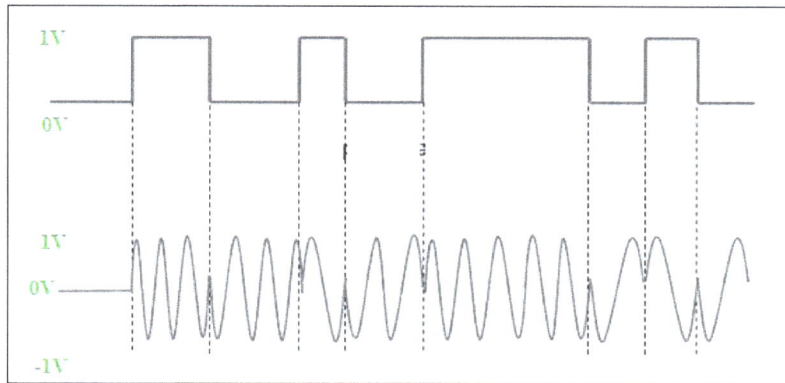

Phase Shift Keying.

M-ary Digital Modulation

In this modulation technique, two or more bits are made for transmitting at once on a single signal, and then it will help in reducing the bandwith. These techniques are classified into three types namely M-ary ASK, M-ary FSK, and M-ary PSK.

Difference between Analog and Digital Modulation

The difference between Analog Modulation and Digital Modulation include the following:

Analog Modulation	Digital Modulation
An AM signal can signify any value in a range.	A DM signal can only signify with a set of discrete values.
In analog modulation (AM), the input must be in the form of analog.	In digital modulation (DM), the input must be the data in the form of digital.
In AM, the value between the max & min is considered to be applicable.	In DM, only two binary numbers are considered applicable such as 1 and 0.
Most of the signals that we transmit in nature are analog like voice signal, and it is much simpler to complete analog modulation compare with digital.	But using digital modulation, you require transmitting through an ADC converter before transmission & a DAC at the end of the receiver for recovering the unique signal. The extra phases required to transmit DM (digital modulation) enhances both the price and difficulty of the transmitter as well as receiver.
The AM can generate a signal to carry the frequently changing data.	The DM generates a signal whose rate changes at particular time intervals.
In AM, it is not easy to disconnect the signal from noise.	In DM, the signal can simply disconnect from noise.

Thus, this is all about an overview of digital modulation methods. This modulation offers more capacity of data, digital data services with compatibility; data security is high, good communication quality, and faster system accessibility. These modulation schemes have a better capacity for conveying a huge amount of data than AM schemes (analog modulation).

Quadrature Amplitude Modulation

Quadrature amplitude modulation (QAM) is the name of a family of digital modulation methods and a related family of analog modulation methods widely used in modern telecommunications to transmit information. It conveys two analog message signals, or two digital bit streams, by changing (modulating) the amplitudes of two carrier waves, using the amplitude-shift keying (ASK) digital modulation scheme or amplitude modulation (AM) analog modulation scheme. The two carrier waves of the same frequency are out of phase with each other by 90°, a condition known as orthogonality and as quadrature. Being the same frequency, the modulated carriers add together, but can be coherently separated (demodulated) because of their orthogonality property. Another key property is that the modulations are low-frequency/low-bandwith waveforms compared to the carrier frequency, which is known as the narrowband assumption.

Phase modulation (analog PM) and phase-shift keying (digital PSK) can be regarded as a special case of QAM, where the magnitude of the modulating signal is a constant, but its sign changes between positive and negative. This can also be extended to frequency modulation (FM) and frequency-shift keying (FSK), for these can be regarded as a special case of phase modulation.

QAM is used extensively as a modulation scheme for digital telecommunication systems, such as in 802.11 Wi-Fi standards. Arbitrarily high spectral efficiencies can be achieved with QAM by setting a suitable constellation size, limited only by the noise level and linearity of the communications channel. QAM is being used in optical fiber systems as bit rates increase; QAM16 and QAM64 can be optically emulated with a 3-path interferometer.

Demodulation of QAM

In a QAM signal, one carrier lags the other by 90°, and its amplitude modulation is customarily referred to as the in-phase component, denoted by $I(t)$. The other modulating function is the quadrature component, $Q(t)$. So the composite waveform is mathematically modeled as,

$$s_s(t) \triangleq \sin(2\pi f_c t)I(t) + \underbrace{\sin\left(2\pi f_c t + \tfrac{\pi}{2}\right)}_{\cos(2\pi f_c t)}Q(t),$$

or

$$s_c(t) \triangleq \cos(2\pi f_c t)I(t) + \underbrace{\cos\left(2\pi f_c t + \tfrac{\pi}{2}\right)}_{-\sin(2\pi f_c t)}Q(t),$$

where f_c is the carrier frequency. At the receiver, a coherent demodulator multiplies the received signal separately with both a cosine and sine signal to produce the received estimates of $I(t)$ and $Q(t)$. For example:

$$r(t) \triangleq s_c(t)\cos(2\pi f_c t) = I(t)\cos(2\pi f_c t)\cos(2\pi f_c t) - Q(t)\sin(2\pi f_c t)\cos(2\pi f_c t).$$

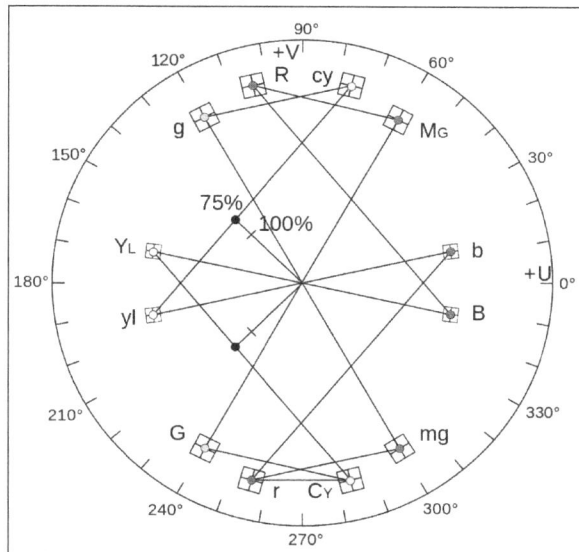

Analog QAM: measured PAL color bar signal on a vector analyzer screen.

Using standard trigonometric identities, we can write this as:

$$r(t) = \frac{1}{2}I(t)[1 + \cos(4\pi fct)] - \frac{1}{2}Q(t)\sin(4\pi fct)$$
$$= \tfrac{1}{2}I(t) + \tfrac{1}{2}[I(t)\cos(4\pi f_c t) - Q(t)\sin(4\pi f_c t)].$$

Low-pass filtering $r(t)$ removes the high frequency terms (containing $4\pi f_c t$), leaving only the $I(t)$ term. This filtered signal is unaffected by $Q(t)$, showing that the in-phase component can be received independently of the quadrature component. Similarly, we can multiply $s_c(t)$ by a sine wave and then low-pass filter to extract $Q(t)$.

The addition of two sinusoids is a linear operation that creates no new frequency components. So the bandwith of the composite signal is comparable to the bandwith of the DSB (Double-Sideband) components. Effectively, the spectral redundancy of DSB enables a doubling of the information capacity using this technique. This comes at the expense of demodulation complexity. In particular, a DSB signal has zero-crossings at a regular frequency, which makes it easy to recover the phase of the carrier sinusoid. It is said to be self-clocking. But the sender and receiver of a quadrature-modulated signal must share a clock or otherwise send a clock signal. If the clock phases drift apart, the demodulated I and Q signals bleed into each other, yielding crosstalk. In this context, the clock signal is called a "phase reference". Clock synchronization is typically achieved by transmitting a burst subcarrier or a pilot signal. The phase reference for NTSC, for example, is included within its color burst signal.

Analog QAM is used in:

- NTSC and PAL analog Color television systems, where the I- and Q-signals carry the components of chroma (colour) information. The QAM carrier phase is recovered from a special Colorburst transmitted at the beginning of each scan line.

- C-QUAM ("Compatible QAM") is used in AM stereo radio to carry the stereo difference information.

Fourier Analysis of QAM

In the frequency domain, QAM has a similar spectral pattern to DSB-SC modulation. Applying Euler's formula to the sinusoids in Eq.1, the positive-frequency portion of s_c (or analytic representation) is:

$$s_c(t)_+ = \tfrac{1}{2}e^{i2\pi f_c t}[I(t) + i \cdot Q(t)] \quad \overset{\mathcal{F}}{\Rightarrow} \quad \tfrac{1}{2}[\widehat{I}(f - f_c) + e^{i\pi/2}\hat{Q}(f - f_c)],$$

where $\mathcal{F}$ denotes the Fourier transform, and $\hat{I}$ and $\hat{Q}$ are the transforms of $I(t)$ and $Q(t)$. This result represents the sum of two DSB-SC signals with the same center frequency. The factor of $i\ (=e^{i\pi/2})$ represents the 90° phase shift that enables their individual demodulations.

Digital QAM

As in many digital modulation schemes, the constellation diagram is useful for QAM. In QAM, the constellation points are usually arranged in a square grid with equal vertical and horizontal spacing, although other configurations are possible (e.g. Cross-QAM). Since in digital telecommunications the data is usually binary, the number of points in the grid is usually a power of 2 (2, 4, 8, ...). Since QAM is usually square, some of these are rare—the most common forms are 16-QAM, 64-QAM and 256-QAM. By moving to a higher-order constellation, it is possible to transmit more bits per symbol. However, if the mean energy of the constellation is to remain the same (by way of making a fair comparison), the points must be closer together and are thus more susceptible to noise and other corruption; this results in a higher bit error rate and so higher-order QAM can deliver more data less reliably than lower-order QAM, for constant mean constellation energy. Using higher-order QAM without increasing the bit error rate requires a higher signal-to-noise ratio (SNR) by increasing signal energy, reducing noise, or both.

If data-rates beyond those offered by 8-PSK are required, it is more usual to move to QAM since it achieves a greater distance between adjacent points in the I-Q plane by distributing the points more evenly. The complicating factor is that the points are no longer all the same amplitude and so the demodulator must now correctly detect both phase and amplitude, rather than just phase.

64-QAM and 256-QAM are often used in digital cable television and cable modem applications. In the United States, 64-QAM and 256-QAM are the mandated modulation schemes for digital cable as standardised by the SCTE in the standard ANSI/SCTE 07 2013. Note that many marketing people will refer to these as QAM-64 and QAM-256. In the UK, 64-QAM is used for digital terrestrial television (Freeview) whilst 256-QAM is used for Freeview-HD.

Communication systems designed to achieve very high levels of spectral efficiency usually employ very dense QAM constellations. For example, current Homeplug AV2 500-Mbit powerline Ethernet devices use 1024-QAM and 4096-QAM, as well as future devices using ITU-T G.hn standard for networking over existing home wiring (coaxial cable, phone lines and power lines); 4096-QAM provides 12 bits/symbol. Another example is ADSL technology for copper twisted pairs, whose constellation size goes up to 32768-QAM (in ADSL terminology this is referred to as bit-loading, or bit per tone, 32768-QAM being equivalent to 15 bits per tone).

Ultra-high capacity Microwave Backhaul Systems also use 1024-QAM. With 1024-QAM, adaptive coding and modulation (ACM) and XPIC, vendors can obtain gigabit capacity in a single 56 MHz channel.

Interference and Noise

In moving to a higher order QAM constellation (higher data rate and mode) in hostile RF/micro-wave QAM application environments, such as in broadcasting or telecommunications, multipath interference typically increases. There is a spreading of the spots in the constellation, decreasing the separation between adjacent states, making it difficult for the receiver to decode the signal appropriately. In other words, there is reduced noise immunity. There are several test parameter measurements which help determine an optimal QAM mode for a specific operating environment. The following three are most significant:

- Carrier/interference ratio,
- Carrier-to-noise ratio,
- Threshold-to-noise ratio.

Information Theory

Information theory is a mathematical approach to the study of coding of information along with the quantification, storage, and communication of information.

Conditions of Occurrence of Events

If we consider an event, there are three conditions of occurrence.

- If the event has not occurred, there is a condition of uncertainty.
- If the event has just occurred, there is a condition of surprise.
- If the event has occurred, a time back, there is a condition of having some information.

These three events occur at different times. The difference in these conditions help us gain knowledge on the probabilities of the occurrence of events

Entropy

When we observe the possibilities of the occurrence of an event, how surprising or uncertain it would be, it means that we are trying to have an idea on the average content of the information from the source of the event.

Entropy can be defined as a measure of the average information content per source symbol. Claude Shannon, the "father of the Information Theory", provided a formula for it as:

$$H = -\sum_i p_i \log_b p_i$$

Where p_i is the probability of the occurrence of character number i from a given stream of characters and b is the base of the algorithm used. Hence, this is also called as Shannon's Entropy.

The amount of uncertainty remaining about the channel input after observing the channel output, is called as Conditional Entropy. It is denoted by $H(x \mid y)$.

Mutual Information

Let us consider a channel whose output is Y and input is X.

Let the entropy for prior uncertainty be X = H(x).

(This is assumed before the input is applied).

To know about the uncertainty of the output, after the input is applied, let us consider Conditional Entropy, given that Y = yk.

$$H(x\,|\,yk) = \sum_{j=0}^{j-1} p(x_j\,|\,y_k) log_2 [\frac{1}{p(x_j\,|\,y_k)}]$$

This is a random variable for,

$H(X\,|\,y = y_0)\ldots\ldots\ldots H(X\,|\,y = y_k)$ with probabilities $p(y_0)\ldots\ldots\ldots p(y_k - 1)$ respectively.

The mean value of $H(x) - H(x\,|\,y)$ for output alphabet y is:

$$H(X\,|\,Y) = \sum_{k=0}^{k-1} H(X\,|\,y = y_k) p(yk)$$

$$= \sum_{k=0}^{k-1}\sum_{j=0}^{j-1} p(x_j\,|\,yk) p(y_k) log_2 [\frac{1}{p(xj\,|\,yk)}]$$

$$= \sum_{k=0}^{k-1}\sum_{j=0}^{j-1} p(x_j, yk) log_2 [\frac{1}{p(xj\,|\,yk)}]$$

Now, considering both the uncertainty conditions (before and after applying the inputs), we come to know that the difference, i.e. $H(x) - H(x\,|\,y)$ must represent the uncertainty about the channel input that is resolved by observing the channel output.

This is called as the Mutual Information of the channel.

Denoting the Mutual Information as $I(x; y)$ we can write the whole thing in an equation, as follows

$$I(x; y) = H(x) - H(x\,|\,y)$$

Hence, this is the equational representation of Mutual Information.

Properties of Mutual Information

These are the properties of Mutual information.

- Mutual information of a channel is symmetric.

$$I(x; y) = I(y; x)$$

- Mutual information is non-negative.

$$I(x;y) \geq 0$$

- Mutual information can be expressed in terms of entropy of the channel output.

$$I(x;y) = H(y) - H(y \mid x)$$

Where $H(y \mid x)$ is a conditional entropy

- Mutual information of a channel is related to the joint entropy of the channel input and the channel output.

$$I(x;y) = H(x) + H(y) - H(x,y)$$

- Where the joint entropy $H(y,x))$ is defined by

$$H(x,y) = \sum_{j=0}^{j-1} \sum_{k=0}^{k-1} p(xj, yk) log_2 (\frac{1}{p(xi, yk)})$$

Channel Capacity

We have so far discussed mutual information. The maximum average mutual information, in an instant of a signaling interval, when transmitted by a discrete memoryless channel, the probabilities of the rate of maximum reliable transmission of data, can be understood as the channel capacity.

It is denoted by C and is measured in bits per channel use.

Discrete Memoryless Source

A source from which the data is being emitted at successive intervals, which is independent of previous values, can be termed as discrete memoryless source.

This source is discrete as it is not considered for a continuous time interval, but at discrete time intervals. This source is memoryless as it is fresh at each instant of time, without considering the previous values.

Source Coding Theorem

The Code produced by a discrete memoryless source, has to be efficiently represented, which is an important problem in communications. For this to happen, there are code words, which represent these source codes.

For example, in telegraphy, we use Morse code, in which the alphabets are denoted by Marks and Spaces. If the letter E is considered, which is mostly used, it is denoted by "." Whereas the letter Q which is rarely used, is denoted by "--.-"

Let us take a look at the block diagram.

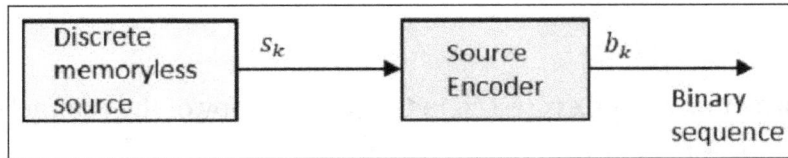

Where Sk is the output of the discrete memoryless source and bk is the output of the source encoder which is represented by 0s and 1s.

The encoded sequence is such that it is conveniently decoded at the receiver.

Let us assume that the source has an alphabet with k different symbols and that the kth symbol S_k occurs with the probability P_k, where k = 0, 1...k-1.

Let the binary code word assigned to symbol S_k by the encoder having length l_k, measured in bits.

Hence, we define the average code word length L of the source encoder as:

$$\overline{L} = \sum_{k=0}^{k-1} p_k l_k$$

$\overline{L}$ represents the average number of bits per source symbol.

If $L_{min} = minimum\ possible\ value\ of\ \overline{L}$

Then coding efficiency can be defined as,

$$\eta = \frac{Lmin}{L}$$

with $\overline{L} \geq L_{min}$ we will have $\eta \leq 1$.

However, the source encoder is considered efficient when $\eta = 1$ For this, the value L_{min} has to be determined.

Let us refer to the definition, "Given a discrete memoryless source of entropy $H(\delta)$ the average code-word length $\overline{L}$ for any source encoding is bounded as $\overline{L} \geq H(\delta)$.

In simpler words, the code word (example: Morse code for the word QUEUE is -.- ..- . ..- .) is always greater than or equal to the source code (QUEUE in example). Which means, the symbols in the code word are greater than or equal to the alphabets in the source code.

Hence with $L_{min} = H(\delta)$ the efficiency of the source encoder in terms of Entropy $H(\delta)$ may be written as:

$$\eta = \frac{H(\delta)}{L}$$

This source coding theorem is called as noiseless coding theorem as it establishes an error-free encoding. It is also called as Shannon's first theorem.

Channel Coding Theorem

The noise present in a channel creates unwanted errors between the input and the output sequences of a digital communication system. The error probability should be very low, nearly $\leq 10^{-6}$ for a reliable communication.

The channel coding in a communication system, introduces redundancy with a control, so as to improve the reliability of the system. The source coding reduces redundancy to improve the efficiency of the system.

Channel coding consists of two parts of action.

- Mapping incoming data sequence into a channel input sequence.

- Inverse Mapping the channel output sequence into an output data sequence.

The final target is that the overall effect of the channel noise should be minimized.

The mapping is done by the transmitter, with the help of an encoder, whereas the inverse mapping is done by the decoder in the receiver.

Channel Coding

Let us consider a discrete memoryless channel (δ) with Entropy H (δ):

T_s indicates the symbols that δ gives per second.

Channel capacity is indicated by C.

Channel can be used for every T_c secs.

Hence, the maximum capability of the channel is C/ T_c.

The data sent $= \dfrac{H(\delta)}{T_s}$.

If $\dfrac{H(\delta)}{T_s} \leq \dfrac{C}{T_c}$ it means the transmission is good and can be reproduced with a small probability of error.

In this, $\dfrac{C}{T_c}$ is the critical rate of channel capacity.

If $\dfrac{H(\delta)}{T_s} = \dfrac{C}{T_c}$ then the system is said to be signaling at a critical rate.

Conversely, if $\dfrac{H(\delta)}{T_s} > \dfrac{C}{T_c}$, then the transmission is not possible.

Hence, the maximum rate of the transmission is equal to the critical rate of the channel capacity, for reliable error-free messages, which can take place, over a discrete memoryless channel. This is called as Channel coding theorem.

Error Control Coding

Noise or Error is the main problem in the signal, which disturbs the reliability of the communication system. Error control coding is the coding procedure done to control the occurrences of errors. These techniques help in Error Detection and Error Correction.

There are many different error correcting codes depending upon the mathematical principles applied to them. But, historically, these codes have been classified into Linear block codes and Convolution codes.

Linear Block Codes

In the linear block codes, the parity bits and message bits have a linear combination, which means that the resultant code word is the linear combination of any two code words.

Let us consider some blocks of data, which contains k bits in each block. These bits are mapped with the blocks which has n bits in each block. Here n is greater than k. The transmitter adds redundant bits which are (n-k) bits. The ratio k/n is the code rate. It is denoted by r and the value of r is $r < 1$.

The (n-k) bits added here, are parity bits. Parity bits help in error detection and error correction, and also in locating the data. In the data being transmitted, the left most bits of the code word correspond to the message bits, and the right most bits of the code word correspond to the parity bits.

Systematic Code

Any linear block code can be a systematic code, until it is altered. Hence, an unaltered block code is called as a systematic code.

Following is the representation of the structure of code word, according to their allocation.

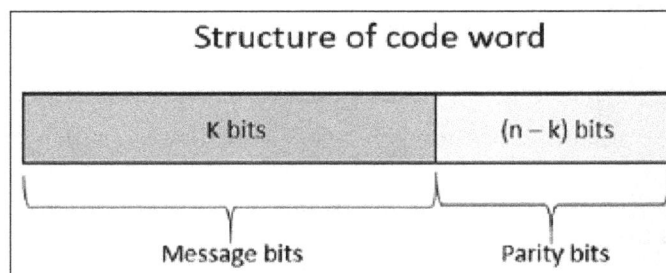

Structure of code word	
K bits	(n – k) bits
Message bits	Parity bits

If the message is not altered, then it is called as systematic code. It means, the encryption of the data should not change the data.

Convolution Codes

So far, in the linear codes, we have discussed that systematic unaltered code is preferred. Here, the data of total n bits if transmitted, k bits are message bits and (n-k) bits are parity bits.

In the process of encoding, the parity bits are subtracted from the whole data and the message bits are encoded. Now, the parity bits are again added and the whole data is again encoded.

The following figure quotes an example for blocks of data and stream of data, used for transmission of information.

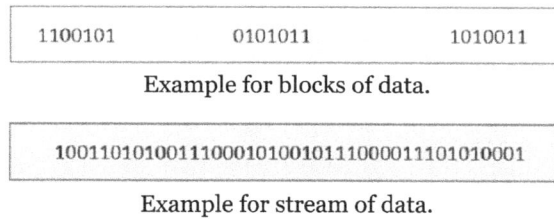

| 1100101 | 0101011 | 1010011 |

Example for blocks of data.

| 10011010100111000101001011100001110101010001 |

Example for stream of data.

The whole process, stated above is tedious which has drawbacks. The allotment of buffer is a main problem here, when the system is busy.

This drawback is cleared in convolution codes. Where the whole stream of data is assigned symbols and then transmitted. As the data is a stream of bits, there is no need of buffer for storage.

Hamming Codes

The linearity property of the code word is that the sum of two code words is also a code word. Hamming codes are the type of linear error correcting codes, which can detect up to two bit errors or they can correct one bit errors without the detection of uncorrected errors.

While using the hamming codes, extra parity bits are used to identify a single bit error. To get from one-bit pattern to the other, few bits are to be changed in the data. Such number of bits can be termed as Hamming distance. If the parity has a distance of 2, one-bit flip can be detected. But this can't be corrected. Also, any two bit flips cannot be detected.

However, Hamming code is a better procedure than the previously discussed ones in error detection and correction.

BCH Codes

BCH codes are named after the inventors Bose, Chaudari and Hocquenghem. During the BCH code design, there is control on the number of symbols to be corrected and hence multiple bit correction is possible. BCH codes is a powerful technique in error correcting codes.

For any positive integers m ≥ 3 and $t < 2^{m-1}$ there exists a BCH binary code. Following are the parameters of such code.

Block length n = 2^m-1.

Number of parity-check digits $n - k \le mt$.

Minimum distance $d_{min} \ge 2t + 1$.

This code can be called as t-error-correcting BCH code.

Cyclic Codes

The cyclic property of code words is that any cyclic-shift of a code word is also a code word. Cyclic codes follow this cyclic property.

For a linear code C, if every code word i.e., C = (C1, C2, ... Cn) from C has a cyclic right shift of components, it becomes a code word. This shift of right is equal to n-1 cyclic left shifts. Hence, it is invariant under any shift. So, the linear code C, as it is invariant under any shift, can be called as a Cyclic code.

Cyclic codes are used for error correction. They are mainly used to correct double errors and burst errors.

Hence, these are a few error correcting codes, which are to be detected at the receiver. These codes prevent the errors from getting introduced and disturb the communication. They also prevent the signal from getting tapped by unwanted receivers.

Spread Spectrum Modulation

A collective class of signaling techniques are employed before transmitting a signal to provide a secure communication, known as the Spread Spectrum Modulation. The main advantage of spread spectrum communication technique is to prevent "interference" whether it is intentional or unintentional.

The signals modulated with these techniques are hard to interfere and cannot be jammed. An intruder with no official access is never allowed to crack them. Hence, these techniques are used for military purposes. These spread spectrum signals transmit at low power density and has a wide spread of signals.

Pseudo-Noise Sequence

A coded sequence of 1s and 0s with certain auto-correlation properties, called as Pseudo-Noise coding sequence is used in spread spectrum techniques. It is a maximum-length sequence, which is a type of cyclic code.

Narrow-band and Spread-spectrum Signals

Both the Narrow band and Spread spectrum signals can be understood easily by observing their frequency spectrum as shown in the following figures.

Narrow-band Signals

The Narrow-band signals have the signal strength concentrated as shown in the following frequency spectrum figure.

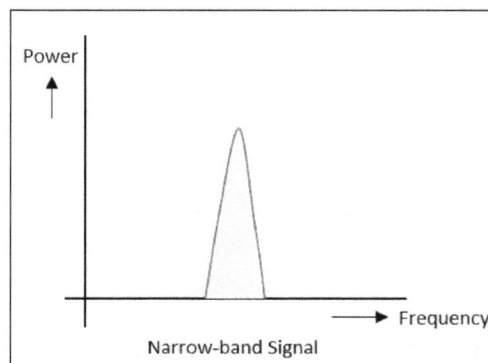

Following are some of its features:

- Band of signals occupy a narrow range of frequencies.
- Power density is high.
- Spread of energy is low and concentrated.

Though the features are good, these signals are prone to interference.

Spread Spectrum Signals

The spread spectrum signals have the signal strength distributed as shown in the following frequency spectrum figure.

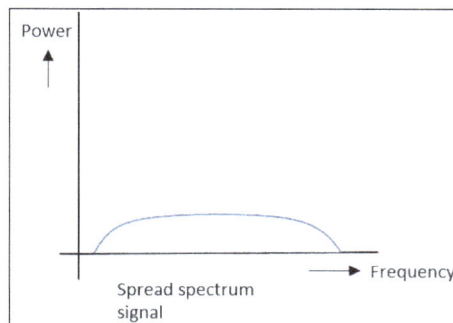

Following are some of its features:

- Band of signals occupy a wide range of frequencies.
- Power density is very low.
- Energy is wide spread.

With these features, the spread spectrum signals are highly resistant to interference or jamming. Since multiple users can share the same spread spectrum bandwith without interfering with one another, these can be called as multiple access techniques.

FHSS and DSSS/CDMA

Spread spectrum multiple access techniques uses signals which have a transmission bandwith of a magnitude greater than the minimum required RF bandwith.

These are of two types:

- Frequency Hopped Spread Spectrum (FHSS).
- Direct Sequence Spread Spectrum (DSSS).

Frequency Hopped Spread Spectrum

This is frequency hopping technique, where the users are made to change the frequencies of usage, from one to another in a specified time interval, hence called as frequency hopping. For example, a frequency was allotted to sender 1 for a particular period of time. Now, after a while, sender 1 hops to the other frequency and sender 2 uses the first frequency, which was previously used by sender 1. This is called as frequency reuse.

The frequencies of the data are hopped from one to another in order to provide a secure transmission. The amount of time spent on each frequency hop is called as Dwell time.

Direct Sequence Spread Spectrum

Whenever a user wants to send data using this DSSS technique, each and every bit of the user data is multiplied by a secret code, called as chipping code. This chipping code is nothing but the spreading code which is multiplied with the original message and transmitted. The receiver uses the same code to retrieve the original message

Comparison between FHSS and DSSS/CDMA

Both the spread spectrum techniques are popular for their characteristics. To have a clear understanding, let us take a look at their comparisons..

FHSS	DSSS/CDMA
Multiple frequencies are used	Single frequency is used
Hard to find the user's frequency at any instant of time	User frequency, once allotted is always the same
Frequency reuse is allowed	Frequency reuse is not allowed
Sender need not wait	Sender has to wait if the spectrum is busy
Power strength of the signal is high	Power strength of the signal is low
Stronger and penetrates through the obstacles	It is weaker compared to FHSS
It is never affected by interference	It can be affected by interference
It is cheaper	It is expensive
This is the commonly used technique	This technique is not frequently used

Advantages of Spread Spectrum

Following are the advantages of spread spectrum:

- Cross-talk elimination.

- Better output with data integrity.

- Reduced effect of multipath fading.

- Better security.

- Reduction in noise.

- Co-existence with other systems.

- Longer operative distances.

- Hard to detect.

- Not easy to demodulate/decode.

- Difficult to jam the signals.

Although spread spectrum techniques were originally designed for military uses, they are now being used widely for commercial purpose.

DIGITAL SIGNAL

A digital signal represents information as a series of binary digits. A binary digit (or bit) can only take one of two values - one or zero. For that reason, the signals used to represent digital information are often waveforms that have only two (or sometimes three) discrete states. In the signal waveform shown below, the signal alternates between two discrete states (0 volts and 5 volts) which could be used to represent binary zero and binary one respectively. If it were actually possible for the signal voltage to instantly transition from zero to five volts (or vice versa), the signal could be said to be discontinuous. In reality, such an instantaneous transition is not physically possible, and a small amount of time is required for the voltage to increase from zero to five volts, and again for the signal to drop from five to zero volts. These finite time periods are referred to as the rise time and the fall time respectively.

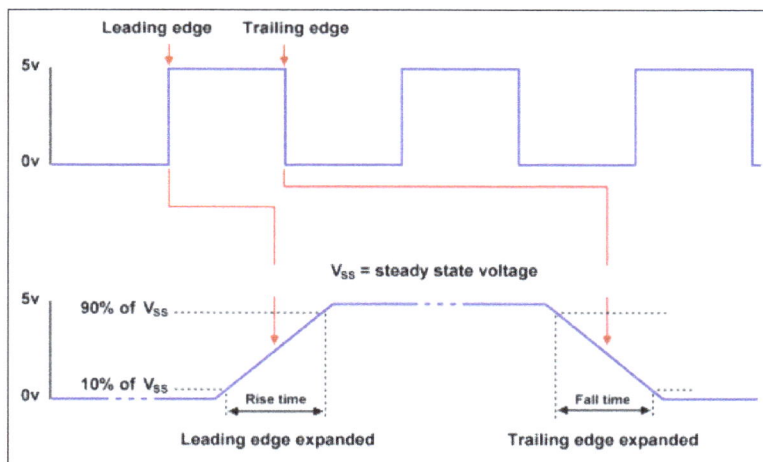

A simple digital signal.

In the simple digital signal represented above, alternating binary ones and zeroes are represented by different voltage levels. A binary one would appear on the transmission line as a short voltage pulse, while a binary zero would be represented as an absence of voltage. This rather simplistic signalling scheme has a number of serious flaws, one of which is that a long series of consecutive ones (or a long series of consecutive zeroes) presents the receiver with the problem of determining exactly how many bits are actually being transmitted. For this to be possible, the duration of each bit-time must be known to both the transmitter and the receiver, and the receiver?s internal clock must be synchronised exactly with that of the transmitter, so that the correct number of consecutive identical bits can be calculated by the receiver. In the example shown below, there are no more than two consecutive bits with the same value, which would not normally present the receiver with too much of a problem. Extended runs of binary numbers having the same value, however, would prove far more of a challenge.

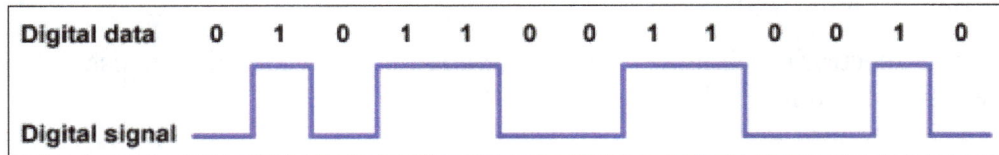

Data representation in a digital signal.

Our simple example in the first diagram uses a positive voltage to represent a one, and the absence of a voltage to represent a zero (for historical reasons, the terms mark and space are often used to refer to the binary digits one and zero respectively). This prompts the question of how the receiver knows whether the transmitter is transmitting a long stream of zeroes, or has simply ceased to transmit. There are, in fact, many different digital encoding schemes that overcome this problem, together with that of long streams of bits having the same value. For now, it is enough to understand that digital signals convey binary data in the form of ones and zeros, using different, discrete signal levels to represent the different logical values. If the signalling scheme used employs a positive voltage to represent one logic state, and a negative voltage to represent the other, the signal is said to be bipolar.

The number of bits that can be transmitted by the signalling scheme in one second is known as its data rate, and is expressed as bits per second (bps), kilobits per second (kbps) or megabits per second (Mbps). The duration of a bit is the time the transmitter takes to output the bit (and as such is obviously related to the data rate). The modulation or signalling rate is the rate at which the signal level is changed, and depends on the digital encoding scheme used (and is also directly related to the data rate). A special case of digital signalling involves the generation of clock signals used to provide synchronisation and timing information for various signal-processing and computing devices. Clock ticks are triggered by either the rising or falling edge (or in some cases both the rising and falling edges) of an alternating digital signal.

The physical communications channel between two communicating end points will inevitably be subject to external noise (electromagnetic interference), so errors will occasionally occur. The degree to which the receiver will be able to correctly interpret incoming signals will depend upon several factors, including its ability to synchronise with the transmitter, the signal-to-noise ratio (SNR), which is a measure of the difference between the transmitted signal strength and the level of background noise, and the data rate. The data rate is significant in this respect because it is directly related to the baseband frequency used. Signals at higher frequencies tend to be more susceptible to very short but high-intensity bursts of external noise (impulse noise), because as frequency increases, there is a greater likelihood that one or more bits in the data stream will become corrupted by a so-called "spike".

In order for the receiver to correctly interpret an incoming stream of bits, it must be able to determine where each bit starts and ends. In order to do this, it needs to somehow be synchronised with the transmitter. It will need to sample each bit as it arrives to determine whether the signal level is high (denoting a binary one) or low (denoting a binary zero). In the simple digital encoding schemes considered so far, each bit will be sampled in the middle of the bit-time, and the measured value compared to pre-determined threshold values to determine whether it is a logic high or a logic low (or neither).

Timing information becomes more critical as data rates increase and the bit duration becomes shorter, especially for data transfers involving large blocks of data consisting of thousands of bits of information. At relatively low data rates, and for asynchronous data transmission involving only a few bits or bytes of data at any one time, the receiver?s internal clock signal will normally suffice to maintain synchronisation with the transmitter long enough to sample the incoming bits in each block of data received at (or close to) the centre of each bit-time (synchronous and asynchronous transmission are dealt with in more detail elsewhere). For larger blocks of data, however, the receiver?s internal clock cannot be relied upon to remain synchronised with the transmitter. A more reliable timing mechanism is required to maintain synchronisation between receiver and transmitter.

One option would be for the transmitter to transmit a separate timing signal which the receiver could use to synchronise its sampling operations on the incoming data stream. This would significantly increase the overall bandwith required for data transmission, and make the digital transmission system far more difficult to design and implement. Fortunately this is not necessary, because the required timing signal can be embedded in the data itself. This is achieved by encoding the data in such a way that there is a guaranteed transition in signal level (from high to low or from low to high) at some point during each bit-time. One such encoding scheme, called Manchester encoding, is illustrated below. This scheme guarantees a transition in the middle of each bit-time that serves as both a clocking mechanism and as a method of encoding the data. A low-to-high transition represents a binary one, while a high-to-low transition represents a binary zero. This type of encoding is known as bi-phase digital encoding. Such schemes are said to be self-clocking, and have no net dc component (there are both positive and negative voltage components of equal duration, during each bit-time).

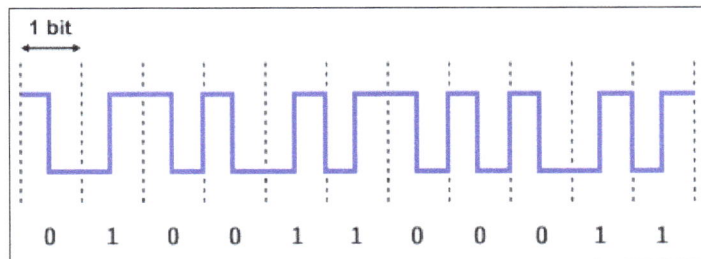

Manchester encoding is a bi-phase digital encoding scheme.

One of the main advantages of digital communications is that virtually any kind of information can be represented digitally, which means that many different kinds of data may be transmitted over the same physical transmission medium. In fact, a number of different digital data streams may share the same physical transmission medium at the same time, thanks to advanced multiplexing techniques. The number of bits required to represent each item of data transmitted will depend on the type of information being sent. Alpha-numeric characters in the ASCII character set, for example, require eight bits per character. Other character encoding schemes can represent a far greater number of characters, but require more bits to represent each character. Analogue information (for example audio or video data) can be represented digitally by sampling the analogue waveform many hundreds, or even thousands of times per second, and then encoding the sample data using a finite range of discrete values (a process known as quantising). The values derived using the quantisation process are then represented as binary numbers, and as such can be transmitted over a digital communications medium as a bit stream. The sampling, quantisation, and conversion to binary format represent an analogue-to-digital conversion (ADC).

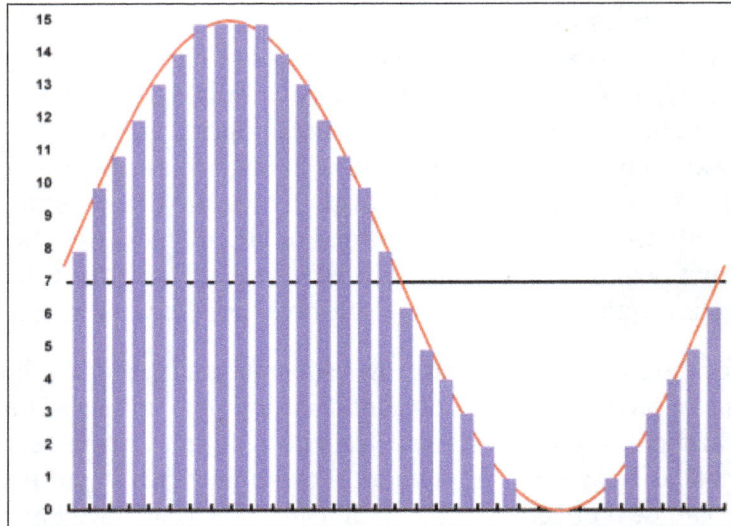

The sampling process repeatedly measures the instantaneous voltage of the analogue waveform.

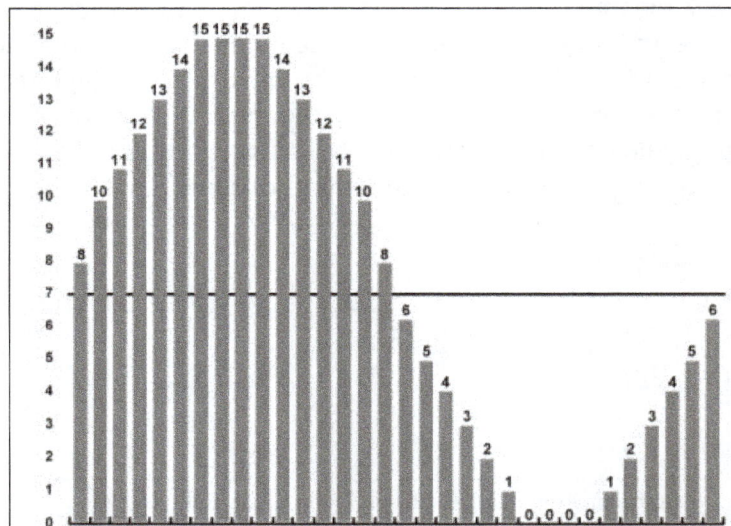

The quantisation process assigns a discrete numeric value to each sample.

```
1000  1010  1011  1100  1101  1110  1111  1111
1111  1111  1110  1101  1100  1011  1010  1000
0110  0101  0100  0011  0010  0001  0000  0000
0000  0000  0001  0010  0011  0100  0101  0110
```

The quantised values are encoded as binary numbers.

The number of bits used to represent each sample will depend on the total number of discrete values required to represent the original data so that the original analogue waveform can be reproduced at the receiver to an acceptable standard. The more samples taken per unit time, the more closely the reconstructed analogue waveform will reflect the original waveform (or, to put it another way, the higher the resolution will be). The cost of higher resolution is that more bits will be required to digitally encode each sample, increasing the bandwith required for transmission. Analogue human voice signals are encoded for transmission over digital circuits in the public switched telephone service (PSTN) using eight bits per sample, giving a range of 256 possible values for each sample. The signals are sampled eight thousand times per second, giving a total requirement of 8 x 8,000 bits per second, or 64 kbps. This is adequate for voice transmission over the telephone network which has traditionally been restricted to a bandwith of less than 4 kHz.

For high-quality real-time video transmission, the data rate (and hence the required transmission bandwith), will be far higher. Various data compression techniques can be used to maximise the bandwith utilisation, but a significant amount of bandwith will still need to be available to guarantee high-quality real-time video transmission, and the complexity of the signal processing required will be greater.

The ability to interleave video, audio, and other forms of data on the same digital transmission links has already been mentioned. Another important advantage of digital signalling is the fact that, because it employs discrete signalling levels, a receiver need only determine whether the sampled voltage represents a logic high (1) or a logic low (0). Small variations in level can otherwise be ignored as having no significance, unlike the continuously varying analogue signals, where even small variations in the amplitude may convey information (or represent fluctuations due to noise). Digital signals suffer from attenuation of course, in the same way that analogue signals suffer from attenuation. Unlike analogue signals, however, as long as a receiver can distinguish between logic high and logic low, the incoming signals can be amplified and repeated with no loss of data whatsoever. The regenerated signal that leaves a digital repeater is identical to the digital signal originally transmitted by the source transmitter.

CHARACTERISTICS OF A DIGITAL SIGNAL

Bit Interval

It is the time required to send one single bit.

Bit Rate

- It refers to the number of bit intervals in one second.

- Therefore bit rate is the number of bits sent in one second as shown in figure.

- Bit rate is expressed in bits per second (bps).

- Other units used to express bit rate are Kbps, Mbps and Gbps.

  1 kilobit per second (Kbps) = 1,000 bits per second.

  1 Megabit per second (Mbps) = 1,000,000 bits per second.

  1 Gigabit per second (Gbps) = 1,000,000,000 bits per second.

NYQUIST–SHANNON SAMPLING THEOREM

In the field of digital signal processing, the sampling theorem is a fundamental bridge between continuous-time signals and discrete-time signals. It establishes a sufficient condition for a sample rate that permits a discrete sequence of *samples* to capture all the information from a continuous-time signal of finite bandwith.

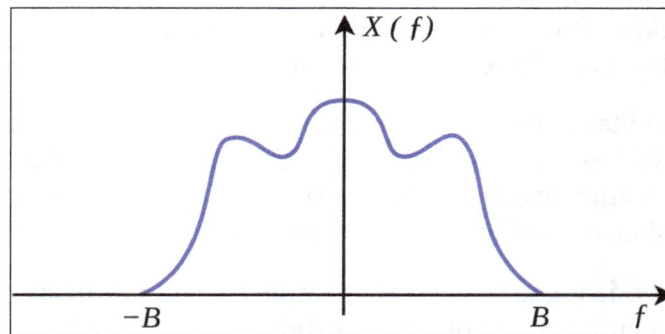

Example of magnitude of the Fourier transform of a bandlimited function.

Strictly speaking, the theorem only applies to a class of mathematical functions having a Fourier transform that is zero outside of a finite region of frequencies. Intuitively we expect that when one reduces a continuous function to a discrete sequence and interpolates back to a continuous function, the fidelity of the result depends on the density (or sample rate) of the original samples. The sampling theorem introduces the concept of a sample rate that is sufficient for perfect fidelity for the class of functions that are bandlimited to a given bandwith, such that no actual information is lost in the sampling process. It expresses the sufficient sample rate in terms of the bandwith for the class of functions. The theorem also leads to a formula for perfectly reconstructing the original continuous-time function from the samples.

Perfect reconstruction may still be possible when the sample-rate criterion is not satisfied, provided other constraints on the signal are known. In some cases (when the sample-rate criterion is not satisfied), utilizing additional constraints allows for approximate reconstructions. The fidelity of these reconstructions can be verified and quantified utilizing Bochner's theorem.

The name Nyquist–Shannon sampling theorem honors Harry Nyquist and Claude Shannon. The theorem was also discovered independently by E. T. Whittaker, by Vladimir Kotelnikov, and by others. It is thus also known by the names Nyquist–Shannon–Kotelnikov, Whittaker–Shannon–Kotelnikov, Whittaker–Nyquist–Kotelnikov–Shannon, and cardinal theorem of interpolation.

Sampling is a process of converting a signal (for example, a function of continuous time and/or space) into a numeric sequence (a function of discrete time and/or space). Shannon's version of the theorem states:

> If a function $x(t)$ contains no frequencies higher than B hertz, it is completely determined by giving its ordinates at a series of points spaced $1/(2B)$ seconds apart.

A sufficient sample-rate is therefore anything larger than $2B$ samples per second. Equivalently, for a given sample rate f_s, perfect reconstruction is guaranteed possible for a bandlimit $B < f_s/2$.

When the bandlimit is too high (or there is no bandlimit), the reconstruction exhibits imperfections known as aliasing. Modern statements of the theorem are sometimes careful to explicitly state that $x(t)$ must contain no sinusoidal component at exactly frequency B, or that B must be strictly less than ½ the sample rate. The threshold $2B$ is called the Nyquist rate and is an attribute of the continuous-time input $x(t)$ to be sampled. The sample rate must exceed the Nyquist rate for the samples to suffice to represent $x(t)$. The threshold $f_s/2$ is called the Nyquist frequency and is an attribute of the sampling equipment. All meaningful frequency components of the properly sampled $x(t)$ exist below the Nyquist frequency. The condition described by these inequalities is called the Nyquist criterion, or sometimes the *Raabe condition*. The theorem is also applicable to functions of other domains, such as *space*, in the case of a digitized image. The only change, in the case of other domains, is the units of measure applied to t, f_s, and B.

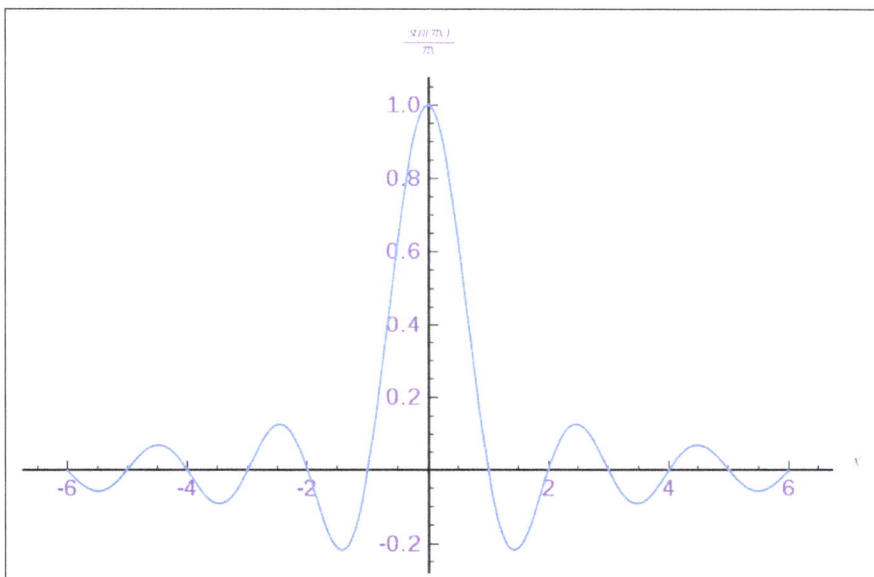

The normalized sinc function: $\sin(\pi x)/(\pi x)$ showing the central peak at $x = 0$, and zero-crossings at the other integer values of x.

The symbol $T = 1/f_s$ is customarily used to represent the interval between samples and is called the sample period or sampling interval. And the samples of function $x(t)$ are commonly denoted by $x[n] = x(nT)$ (alternatively "x_n" in older signal processing literature), for all integer values of n. A mathematically ideal way to interpolate the sequence involves the use of sinc functions. Each sample in the sequence is replaced by a sinc function, centered on the time axis at the original location of the sample, nT, with the amplitude of the sinc function scaled to the sample value, $x[n]$. Subsequently, the sinc functions are summed into a continuous function. A mathematically equivalent method is to convolve one sinc function with a series of Dirac delta pulses, weighted by the sample values. Neither method is numerically practical. Instead, some type of approximation of the sinc functions, finite in length, is used. The imperfections attributable to the approximation are known as *interpolation error*.

Practical digital-to-analog converters produce neither scaled and delayed sinc functions, nor ideal Dirac pulses. Instead they produce a piecewise-constant sequence of scaled and delayed rectangular pulses (the zero-order hold), usually followed by an "anti-aliasing filter" to clean up spurious high-frequency content.

Aliasing

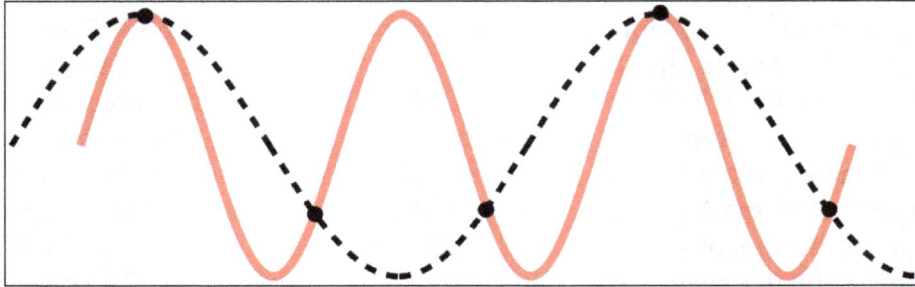

The samples of two sine waves can be identical when at least one of them.
is at a frequency above half the sample rate.

When $x(t)$ is a function with a Fourier transform $X(f)$:

$$X(f) \triangleq \int_{-\infty}^{\infty} x(t)\, e^{-i2\pi ft} \, \mathrm{d}t.$$

The Poisson summation formula indicates that the samples, $x(nT)$, of $x(t)$ are sufficient to create a periodic summation of $X(f)$. The result is:

$$X_s(f) \triangleq \sum_{k=-\infty}^{\infty} X\left(f - kf_s\right) = \sum_{n=-\infty}^{\infty} T \cdot x(nT)\, e^{-i2\pi nTf}$$

which is a periodic function and its equivalent representation as a Fourier series, whose coefficients are $T \cdot x(nT)$. This function is also known as the discrete-time Fourier transform (DTFT) of the sample sequence.

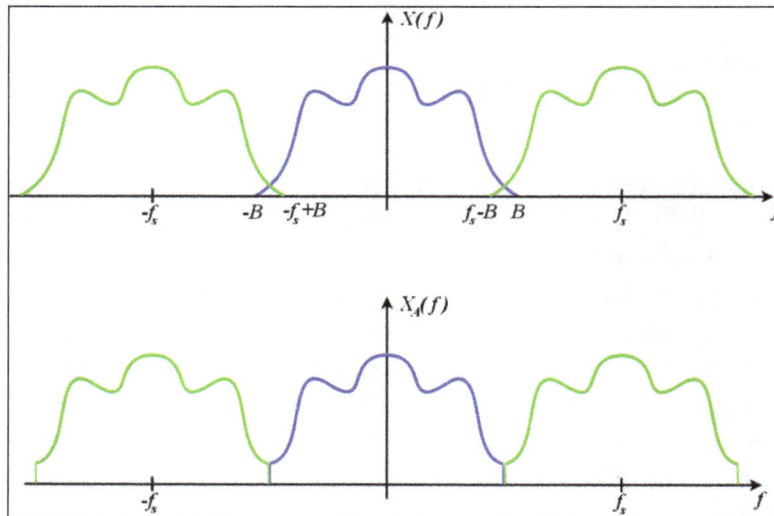

$X(f)$ (top blue) and $X_A(f)$ (bottom blue) are continuous Fourier transforms of two *different* functions, $x(t)$ and $x_A(t)$ (not shown). When the functions are sampled at rate f_s, the images (green) are added to the original transforms (blue) when one examines the discrete-time Fourier transforms (DTFT) of the sequences. In this hypothetical example, the DTFTs are identical, which means *the sampled sequences are identical*, even though the original continuous pre-sampled functions are not. If these were audio signals, $x(t)$ and $x_A(t)$ might not sound the same. But their samples (taken at rate f_s) are identical and would lead to identical reproduced sounds; thus $x_A(t)$ is an alias of $x(t)$ at this sample rate.

As depicted, copies of $X(f)$ are shifted by multiples of f_s and combined by addition. For a band-limited function ($X(f) = 0$, for all $|f| \geq B$) and sufficiently large f_s, it is possible for the copies to remain distinct from each other. But if the NY Quist criterion is not satisfied, adjacent copies overlap, and it is not possible in general to discern an unambiguous $X(f)$. Any frequency component above $f_s / 2$ is indistinguishable from a lower-frequency component, called an *alias*, associated with one of the copies. In such cases, the customary interpolation techniques produce the alias, rather than the original component. When the sample-rate is pre-determined by other considerations (such as an industry standard), $x(t)$ is usually filtered to reduce its high frequencies to acceptable levels before it is sampled. The type of filter required is a lowpass filter, and in this application it is called an anti-aliasing filter.

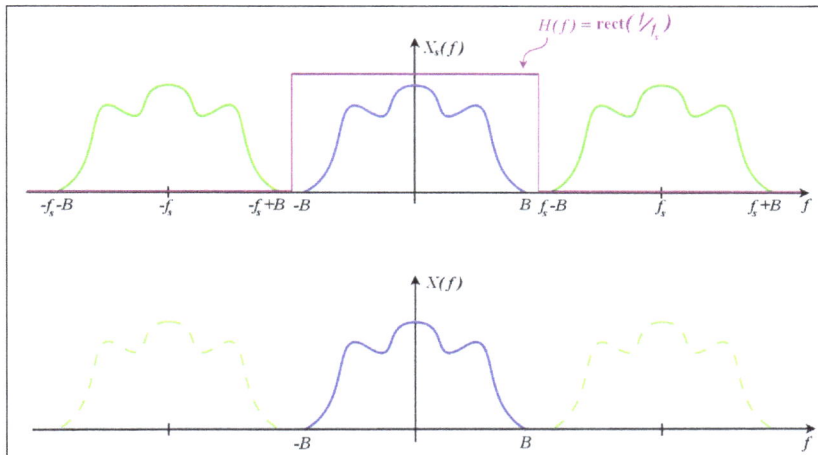

Spectrum, $X_s(f)$, of a properly sampled bandlimited signal (blue) and the adjacent DTFT images (green) that do not overlap. A *brick-wall* low-pass filter, $H(f)$, removes the images, leaves the original spectrum, $X(f)$, and recovers the original signal from its samples.

Derivation as a Special case of Poisson Summation

When there is no overlap of the copies (also known as "images") of , the term of $X_s(f) \triangleq \sum_{k=-\infty}^{\infty} X\left(f - kf_s\right) = \sum_{n=-\infty}^{\infty} T \cdot x(nT)\, e^{-i2\pi nTf}$, can be recovered by the product,

$$X(f) = H(f) \cdot X_s(f),$$

where,

$$H(f) \triangleq \begin{cases} 1 & |f| < B \\ 0 & |f| > f_s - B. \end{cases}$$

At this point, the sampling theorem is proved, since $X(f)$ uniquely determines $x(t)$.

All that remains is to derive the formula for reconstruction. $H(f)$ need not be precisely defined in the region $[B, f_s - B]$ because $X_s(f)$ is zero in that region. However, the worst case is when $B = f_s / 2$, the Nyquist frequency. A function that is sufficient for that and all less severe cases is,

$$H(f) = \text{rect}\left(\frac{f}{f_s}\right) = \begin{cases} 1 & |f| < \dfrac{f_s}{2} \\ 0 & |f| > \dfrac{f_s}{2}, \end{cases}$$

where rect($\cdot$) is the rectangular function. Therefore,

$$\begin{aligned} X(f) &= \text{rect}\left(\frac{f}{f_s}\right) \cdot X_s(f) \\ &= \text{rect}(Tf) \cdot \sum_{n=-\infty}^{\infty} T \cdot x(nT)\, e^{-i2\pi nTf} \\ &= \sum_{n=-\infty}^{\infty} x(nT) \cdot \underbrace{T \cdot \text{rect}(Tf) \cdot e^{-i2\pi nTf}}_{\mathcal{F}\left\{\text{sinc}\left(\frac{t-nT}{T}\right)\right\}}. \end{aligned}$$

The inverse transform of both sides produces the Whittaker–Shannon interpolation formula,

$$x(t) = \sum_{n=-\infty}^{\infty} x(nT) \cdot \text{sinc}\left(\frac{t-nT}{T}\right),$$

which shows how the samples, $x(nT)$, can be combined to reconstruct $x(t)$.

- Larger-than-necessary values of f_s (smaller values of T), called *oversampling*, have no effect on the outcome of the reconstruction and have the benefit of leaving room for a *transition band* in which $H(f)$ is free to take intermediate values. Undersampling, which causes aliasing, is not in general a reversible operation.

- Theoretically, the interpolation formula can be implemented as a low pass filter, whose impulse response is sinc(t/T) and whose input is $\sum_{n=-\infty}^{\infty} x(nT) \cdot \delta(t-nT)$, which is a Dirac comb function modulated by the signal samples. Practical digital-to-analog converters (DAC) implement an approximation like the zero-order hold. In that case, oversampling can reduce the approximation error.

Shannon's Original Proof

Poisson shows that the Fourier series in Eq.1 produces the periodic summation of , regardless of f_s and B. Shannon, however, only derives the series coefficients for the case $f_s = 2B$.

Let $X(\omega)$ be the spectrum of $x(t)$. Then,

$$x(t) = \frac{1}{2\pi} \int_{-\infty}^{\infty} X(\omega)e^{i\omega t}\, d\omega = \frac{1}{2\pi} \int_{-2\pi B}^{2\pi B} X(\omega)e^{i\omega t}\, d\omega,$$

because $X(\omega)$ is assumed to be zero outside the band $\left|\frac{\omega}{2\pi}\right| < B$. If we let $t = \frac{n}{2B}$, where n is any positive or negative integer, we obtain:

$$x\left(\tfrac{n}{2B}\right) = \frac{1}{2\pi} \int_{-2\pi B}^{2\pi B} X(\omega) e^{i\omega \frac{n}{2B}} \, d\omega.$$

On the left are values of $x(t)$ at the sampling points. The integral on the right will be recognized as essentially the nth coefficient in a Fourier-series expansion of the function $X(\omega)$, taking the interval $-B$ to B as a fundamental period. This means that the values of the samples $x(n/2B)$ determine the Fourier coefficients in the series expansion of $X(\omega)$. Thus they determine $X(\omega)$ since $X(\omega)$ is zero for frequencies greater than B, and for lower frequencies $X(\omega)$ is determined if its Fourier coefficients are determined. But determines the original function $x(t)$ completely, since a function is determined if its spectrum is known. Therefore the original samples determine the function $x(t)$ completely.

Shannon's proof of the theorem is complete at that point, but he goes on to discuss reconstruction via sinc functions, what we now call the Whittaker–Shannon interpolation formula as discussed above. He does not derive or prove the properties of the sinc function, but these would have been-familiar to engineers reading his works at the time, since the Fourier pair relationship between rect (the rectangular function) and sinc was well known.

Let x_n be the nth sample. Then the function $x(t)$ is represented by:

$$x(t) = \sum_{n=-\infty}^{\infty} x_n \frac{\sin \pi(2Bt - n)}{\pi(2Bt - n)}.$$

As in the other proof, the existence of the Fourier transform of the original signal is assumed, so the proof does not say whether the sampling theorem extends to bandlimited stationary random processes.

The actual coefficient formula contains an additional factor of $1/2B = T$. So Shannon's coefficients are $T \cdot x(nT)$, which agrees with $X_s(f) \triangleq \sum_{k=-\infty}^{\infty} X(f - kf_s) = \sum_{n=-\infty}^{\infty} T \cdot x(nT) e^{-i2\pi nTf}$.

Application to Multivariable Signals and Images

Subsampled image showing a Moiré pattern.   Properly sampled image.

The sampling theorem is usually formulated for functions of a single variable. Consequently, the theorem is directly applicable to time-dependent signals and is normally formulated in that context. However, the sampling theorem can be extended in a straightforward way to functions of arbitrarily many variables. Grayscale images, for example, are often represented as two-dimensional arrays (or matrices) of real numbers representing the relative intensities of pixels (picture elements) located at the intersections of row and column sample locations. As a result, images require two independent variables, or indices, to specify each pixel uniquely—one for the row, and one for the column.

Color images typically consist of a composite of three separate grayscale images, one to represent each of the three primary colors—red, green, and blue, or *RGB* for short. Other colorspaces using 3-vectors for colors include HSV, CIELAB, XYZ, etc. Some colorspaces such as cyan, magenta, yellow, and black (CMYK) may represent color by four dimensions. All of these are treated as vector-valued functions over a two-dimensional sampled domain.

Similar to one-dimensional discrete-time signals, images can also suffer from aliasing if the sampling resolution, or pixel density, is inadequate. For example, a digital photograph of a striped shirt with high frequencies (in other words, the distance between the stripes is small), can cause aliasing of the shirt when it is sampled by the camera's image sensor. The aliasing appears as a moiré pattern. The "solution" to higher sampling in the spatial domain for this case would be to move closer to the shirt, use a higher resolution sensor, or to optically blur the image before acquiring it with the sensor.

Another example is shown to the right in the brick patterns. The top image shows the effects when the sampling theorem's condition is not satisfied. When software rescales an image (the same process that creates the thumbnail shown in the lower image) it, in effect, runs the image through a low-pass filter first and then downsamples the image to result in a smaller image that does not exhibit the moiré pattern. The top image is what happens when the image is downsampled without low-pass filtering: aliasing results.

The sampling theorem applies to camera systems, where the scene and lens constitute an analog spatial signal source, and the image sensor is a spatial sampling device. Each of these components is characterized by a modulation transfer function (MTF), representing the precise resolution (spatial bandwith) available in that component. Effects of aliasing or blurring can occur when the lens MTF and sensor MTF are mismatched. When the optical image which is sampled by the sensor device contains higher spatial frequencies than the sensor, the under sampling acts as a low-pass filter to reduce or eliminate aliasing. When the area of the sampling spot (the size of the pixel sensor) is not large enough to provide sufficient spatial anti-aliasing, a separate anti-aliasing filter (optical low-pass filter) may be included in a camera system to reduce the MTF of the optical image. Instead of requiring an optical filter, the graphics processing unit of smartphone cameras performs digital signal processing to remove aliasing with a digital filter. Digital filters also apply sharpening to amplify the contrast from the lens at high spatial frequencies, which otherwise falls off rapidly at diffraction limits.

The sampling theorem also applies to post-processing digital images, such as to up or down sampling. Effects of aliasing, blurring, and sharpening may be adjusted with digital filtering implemented in software, which necessarily follows the theoretical principles.

Critical Frequency

To illustrate the necessity of $f_s > 2B$, consider the family of sinusoids generated by different values of θ in this formula:

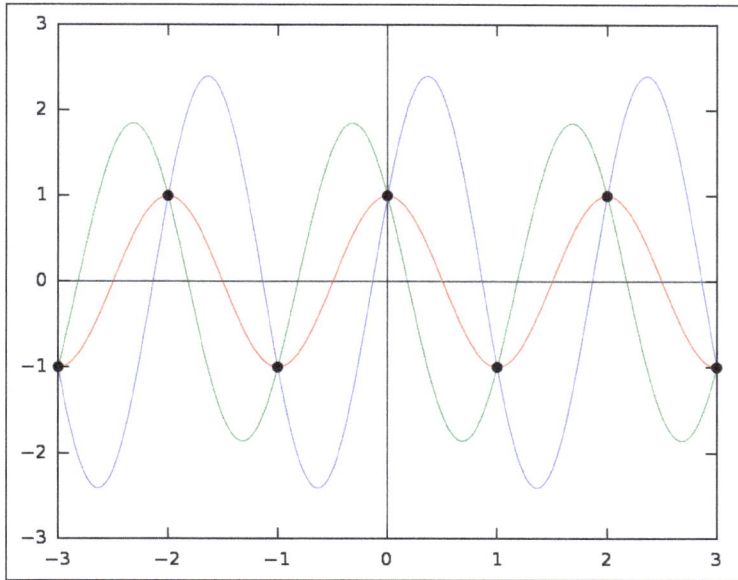

A family of sinusoids at the critical frequency, all having the same sample
sequences of alternating +1 and −1. That is, they all are aliases of each other,
even though their frequency is not above half the sample rate.

$$x(t) = \frac{\cos(2\pi Bt + \theta)}{\cos(\theta)} = \cos(2\pi Bt) - \sin(2\pi Bt)\tan(\theta), \quad -\pi/2 < \theta < \pi/2.$$

With $f_s = 2B$ or equivalently $T = 1/2B$, the samples are given by,

$$x(nT) = \cos(\pi n) - \underbrace{\sin(\pi n)}_{0}\tan(\theta) = (-1)^n$$

regardless of the value of θ. That sort of ambiguity is the reason for the *strict* inequality of the sampling theorem's condition.

Sampling of Non-baseband Signals

As discussed by Shannon:

> A similar result is true if the band does not start at zero frequency but at some higher value, and can be proved by a linear translation (corresponding physically to single-sideband modulation) of the zero-frequency case. In this case the elementary pulse is obtained from $\sin(x)/x$ by single-side-band modulation.

That is, a sufficient no-loss condition for sampling signals that do not have baseband components exists that involves the *width* of the non-zero frequency interval as opposed to its highest frequency component.

For example, in order to sample the FM radio signals in the frequency range of 100–102 MHz, it is not necessary to sample at 204 MHz (twice the upper frequency), but rather it is sufficient to sample at 4 MHz (twice the width of the frequency interval).

A bandpass condition is that $X(f) = 0$, for all nonnegative f outside the open band of frequencies:

$$\left(\frac{N}{2} f_s, \frac{N+1}{2} f_s \right),$$

for some nonnegative integer N. This formulation includes the normal baseband condition as the case $N=0$.

The corresponding interpolation function is the impulse response of an ideal brick-wall bandpass filter (as opposed to the ideal brick-wall lowpass filter used above) with cutoffs at the upper and lower edges of the specified band, which is the difference between a pair of lowpass impulse responses:

$$(N+1)\operatorname{sinc}\left(\frac{(N+1)t}{T} \right) - N\operatorname{sinc}\left(\frac{Nt}{T} \right).$$

Other generalizations, for example to signals occupying multiple non-contiguous bands, are possible as well. Even the most generalized form of the sampling theorem does not have a provably true converse. That is, one cannot conclude that information is necessarily lost just because the conditions of the sampling theorem are not satisfied; from an engineering perspective, however, it is generally safe to assume that if the sampling theorem is not satisfied then information will most likely be lost.

Nonuniform Sampling

The sampling theory of Shannon can be generalized for the case of nonuniform sampling, that is, samples not taken equally spaced in time. The Shannon sampling theory for non-uniform sampling states that a band-limited signal can be perfectly reconstructed from its samples if the average sampling rate satisfies the Nyquist condition. Therefore, although uniformly spaced samples may result in easier reconstruction algorithms, it is not a necessary condition for perfect reconstruction.

The general theory for non-baseband and nonuniform samples was developed in 1967 by Henry Landau. He proved that the average sampling rate (uniform or otherwise) must be twice the *occupied* bandwith of the signal, assuming it is *a priori* known what portion of the spectrum was occupied. In the late 1990s, this work was partially extended to cover signals of when the amount of occupied bandwith was known, but the actual occupied portion of the spectrum was unknown. In the 2000s, a complete theory was developed using compressed sensing. In particular, the theory, using signal processing language, is described in this 2009 paper. They show, among other things, that if the frequency locations are unknown, then it is necessary to sample at least at twice the Nyquist criteria; in other words, you must pay at least a factor of 2 for not knowing the location of the spectrum. Note that minimum sampling requirements do not necessarily guarantee stability.

Sampling below the Nyquist Rate under Additional Restrictions

The Nyquist–Shannon sampling theorem provides a sufficient condition for the sampling and reconstruction of a band-limited signal. When reconstruction is done via the Whittaker–Shannon interpolation formula, the Nyquist criterion is also a necessary condition to avoid aliasing, in the sense that if samples are taken at a slower rate than twice the band limit, then there are some signals that will not be correctly reconstructed. However, if further restrictions are imposed on the signal, then the Nyquist criterion may no longer be a necessary condition.

A non-trivial example of exploiting extra assumptions about the signal is given by the recent field of compressed sensing, which allows for full reconstruction with a sub-Nyquist sampling rate. Specifically, this applies to signals that are sparse (or compressible) in some domain. As an example, compressed sensing deals with signals that may have a low over-all bandwith (say, the *effective* bandwith EB), but the frequency locations are unknown, rather than all together in a single band, so that the passband technique does not apply. In other words, the frequency spectrum is sparse. Traditionally, the necessary sampling rate is thus $2B$. Using compressed sensing techniques, the signal could be perfectly reconstructed if it is sampled at a rate slightly lower than $2EB$. With this approach, reconstruction is no longer given by a formula, but instead by the solution to a linear optimization program.

Another example where sub-Nyquist sampling is optimal arises under the additional constraint that the samples are quantized in an optimal manner, as in a combined system of sampling and optimal lossy compression. This setting is relevant in cases where the joint effect of sampling and quantization is to be considered, and can provide a lower bound for the minimal reconstruction error that can be attained in sampling and quantizing a random signal. For stationary Gaussian random signals, this lower bound is usually attained at a sub-Nyquist sampling rate, indicating that sub-Nyquist sampling is optimal for this signal model under optimal quantization.

References

- Digital-communication-analog-to-digital, digital-communication: tutorialspoint.com, Retrieved 2 June, 2019

- Digital-communication-sampling, digital-communication: tutorialspoint.com, Retrieved 22 May, 2019

- Sayood, Khalid (2005), Introduction to Data Compression, Third Edition, Morgan Kaufmann, ISBN 978-0-12-620862-7

- Digital-communication-differential-pcm, digital-communication: tutorialspoint.com, Retrieved 23 February, 2019

- Digital-communication-techniques, digital-communication: tutorialspoint.com, Retrieved 13 March, 2019

- Digital-modulation-different-types-and-their-differences: elprocus.com, Retrieved 29 June, 2019

- "Ciena tests 200G via 16-QAM with Japan-U.S. Cable Network". Lightwave. April 17, 2014. Retrieved 7 November 2016

- Digital-communication-information-theory, digital-communication: tutorialspoint.com, Retrieved 30 January, 2019

- Digital-signal, communication-networks, computernetworkingnotes: ecomputernotes.com, Retrieved 31 March, 2019

Pulse Modulation

Pulse modulation is a technique where signal is transmitted along with the information by pulses. It is categorized into pulse-width modulation, pulse position modulation, pulse code modulation and delta modulation. This chapter discusses in detail these different methods and techniques related to pulse modulation.

Pulse modulation is a technique in which the signal is transmitted with the information by pulses. This is divided into Analog Pulse Modulation and Digital Pulse Modulation.

Analog pulse modulation is classified as:

- Pulse Amplitude Modulation (PAM).

- Pulse Width Modulation (PWM).

- Pulse Position Modulation (PPM).

Digital modulation is classified as:

- Pulse Code Modulation.

- Delta Modulation.

Pulse Amplitude Modulation

Pulse amplitude modulation is a technique in which the amplitude of each pulse is controlled by the instantaneous amplitude of the modulation signal. It is a modulation system in which the signal is sampled at regular intervals and each sample is made proportional to the amplitude of the signal at the instant of sampling. This technique transmits the data by encoding in the amplitude of a series of signal pulses.

There are two types of sampling techniques for transmitting a signal using PAM. They are:

- Flat Top PAM.

- Natural PAM.

Flat Top PAM: The amplitude of each pulse is directly proportional to modulating signal amplitude at the time of pulse occurrence. The amplitude of the signal cannot be changed with respect to the analog signal to be sampled. The tops of the amplitude remain flat.

Natural PAM: The amplitude of each pulse is directly proportional to modulating signal amplitude at the time of pulse occurrence. Then follows the amplitude of the pulse for the rest of the half cycle.

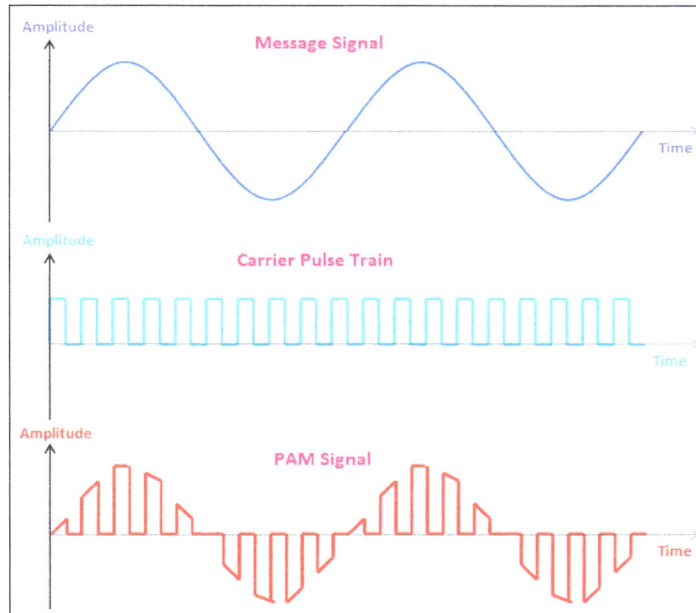

Pulse Amplitude Modulation Signal.

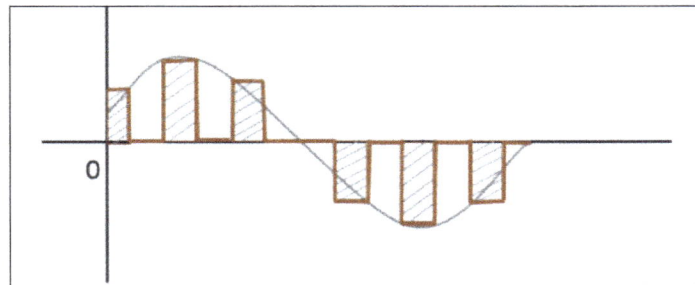

Flat Top PAM.

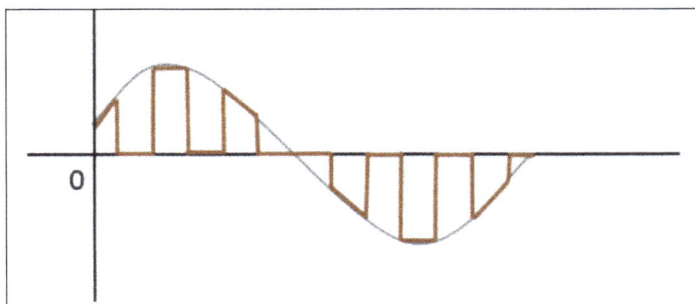

Natural PAM.

Circuit Design of Pulse Amplitude Modulation

PAM is generated from a pure sine wave modulating signal and a square wave generator which produces the carrier pulse and a PAM modulator circuit.

A sine wave generator is used which is based on Wien Bridge Oscillator circuit. This can produce distortion less sine wave at the output. The circuit is designed such that the amplitude and the frequency of the oscillator can be adjusted using a potentiometer.

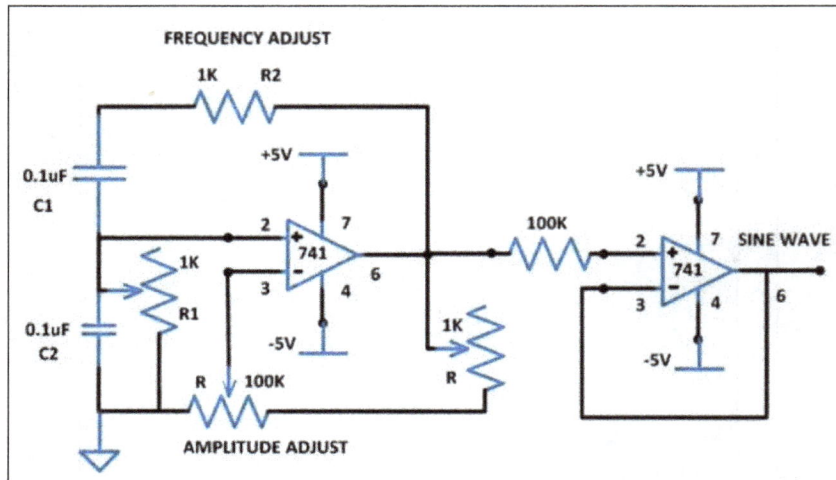

Sine Wave Generator.

The frequency can be varied by varying the potentiometer R2 and the amplitude of the adjusted using the potentiometer R. The frequency of the sine wave generated is given by,

$$F = 1/(2\pi\sqrt{R1R_2C1C_2})$$

The square wave is generated using op-amp based astable circuit. The op-amp is used to reduce the complexity of generating the square wave. The ON time and the OFF time of the pulse can be made identical and the frequency can be adjusted without changing them.

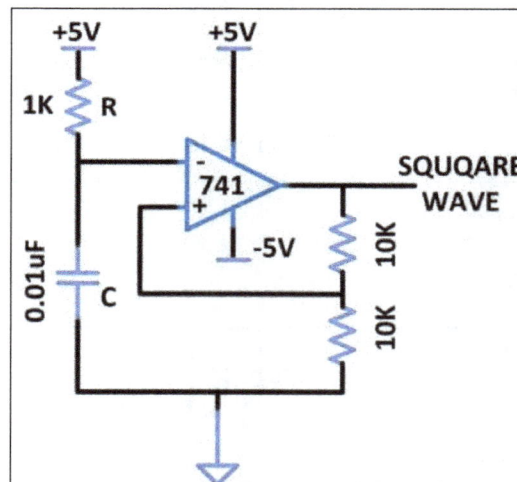

Square Wave Generator.

The time period of the pulses generated depends on the value of the resistance R and the capacitance C. The period of the op-amp astable circuit is given by,

$$T = 2.2RC$$

Pulse Width Modulation

Pulse width modulation is that technique by which a low-frequency signal is being generated from high-frequency pulses. The main use of pulse width modulation is to let the higher load electrical devices to take over the control of the power supplied to the system. In a simpler sense, pulse width modulation uses digital signals to control high power applications. Besides, it is easy to convert the PWM signals back to the analog form with the minimum use of hardware. The power is controlled by turning a switch on and off between the supply and load at an increased rate. The power supplied to the load is directly proportional to the duration of the on time. Pulse width modulation is also called pulse duration modulation.

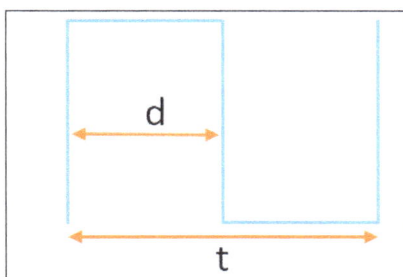

The proportion of 'on' time to a period of time is called duty cycle. Duty cycle is expressed in percentage. Duty cycle is the inverse of the frequency of the waveform. The switching frequency of pulse width modulation must be higher than the rate which would affect the power device. This frequency can be different for different applications and devices.

By imparting pulse width modulation in an application, the power loss in the switching device can be considerably reduced. When the switch is made on, the voltage drop across the switch is practically zero. And when the switch is off, there is no current. In this case, also power loss is negligible, as the power is the product of voltage and current.

The on/off nature of pulse width modulation is another advantage because of which the PWM is widely used in various digital control applications. It is easier with a PWM to set the required duty cycle. Because of this property, pulse width modulation has found its applications in communication systems as well.

The heat produced during the operation of a digital system is lesser compared to that during the working of an analog system. Majority of the heat in a switching device is generated during the transition phase. The device is at a state between on and off at this time. This is because power is the product of voltage and current. During the working of pulse width modulation system, either current or voltage is nearly zero. And therefore the heat produced is almost zero in such systems.

A pulse width modulation signal can be generated by a sawtooth waveform as well as using a comparator.

Basic Principle

Pulse width modulation makes use of a rectangular pulse wave of which the pulse width is modulated. This modulation of pulse width results in the variation of the average value of the waveform, where the average value is dependent on the duty cycle D.

Duty cycle is the period in which the signal is on and is denoted as a percentage. Duty cycle is 50 percent when the signal is 'on' for half of the time period.

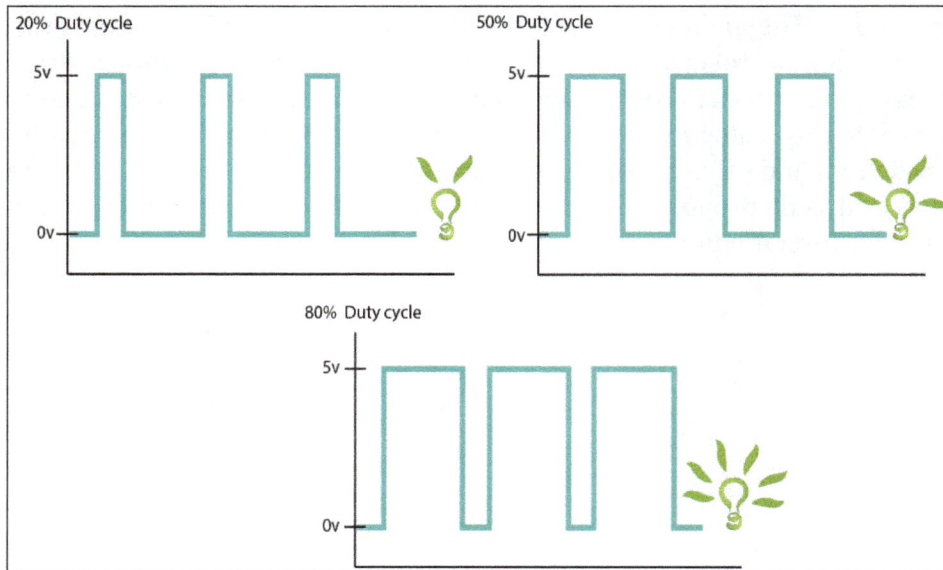

Consider the case of 20% duty cycle for a voltage switching between 0V and 5V, the average voltage will be 5×0.2=1V. Similarly, for 50% and 80% duty cycle it is 2.5V and 4V respectively. This property of PWM is very useful in motor speed control and LED brightness control.

PWM Generation in Analog Circuits

To generate PWM signals in analog circuits, a sawtooth triangle waveform is used. It can be easily generated using a simple oscillator. A comparator can also be included. The logic of generating PWM signals is very simple. If the value of the reference signal (sawtooth triangle signal) is more than the modulation waveform, the resultant PWM signal is a high signal, otherwise, the resultant waveform is in the low state. This method is also called as a carrier-based generation where the reference signal act as the carrier waveform.

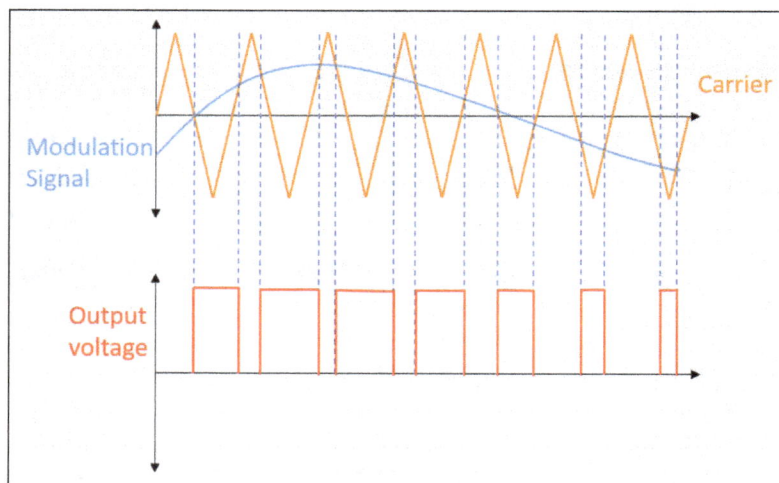

PWM Generation in Digital Circuits

Most of the day-to-day digital circuits can produce a pulse width modulation signal. for example, microcontrollers generate PWM output signals. In microcontrollers, in order to generate a pulse width modulation signal, a counter which increments periodically is used. This counter is resent at the end of every period of the PWM. The PWM output changes state from high to low when the counter value is more than the reference value. This technique is called time proportioning or time proportioning control.

This counter is considered as the counterpart of the reference sawtooth signal in the analog circuits. The output PWM waveform is generated by comparing the current counter value and the digital reference value. The duty cycle varies in discrete steps. As the resolution of the counter increases, the performance of the counter becomes better.

Applications

One of the very popular applications of pulse width modulation is LED lighting. Usually, an LED has the tendency to produce non-linear light by applying a current. By applying the pulse width modulation, one can produce a linear light output. The brightness of the LED can be controlled by adjusting the duty cycle. In a Red-Green-Blue LED system (RGB), if the brightness of all the three LEDs is equal, then the resultant output light will be white. By varying the duty cycle of various LEDs, different colors are produced.

Another important application of pulse width modulation signal is a motor. Pulse width modulation is used in a servo motor to control the angle of the motor which is attached to a robot arm or similar things. The shaft attached to the servo motor turns to specific positions based on the control line. There are some microcontrollers with built-in PWM.

Pulse width modulation is highly useful in motor speed control applications. By incorporating pulse width modulation, it becomes quite easier for the motor to vary the speed. And this increases the efficiency of the system. It is more effective to control the motor using PWM at low RPM than linear methods. H-Bridge makes use of the pulse width modulation in an effective manner. Since the power supply can be switched across both sides of the load, the effective voltage across the load will be doubled.

In a class-D audio amplifier, the maximum output is increased by increasing the voltage. PWM is the pinnacle of such amplifiers and is used by choosing a frequency beyond human hearing. Pulse width modulation is being used by switching mode power supplies.

PWM is also used in switched mode power supplies (SMPS). By switching voltage to the load in a controlled manner with respect to duty cycle, the voltage can be maintained at the desired level. By using the feedback, the load can be monitored and power supply becomes much more efficient.

PULSE POSITION MODULATION

Pulse-position modulation (PPM) is a form of signal modulation in which M message bits are encoded by transmitting a single pulse in one of 2^M possible required time shifts. This is repeated every T seconds, such that the transmitted bit rate is M/T bits per second. It is primarily useful for optical communications systems, which tend to have little or no multipath interference.

Synchronization

One of the key difficulties of implementing this technique is that the receiver must be properly synchronized to align the local clock with the beginning of each symbol. Therefore, it is often implemented differentially as *differential pulse-position modulation*, whereby each pulse position is encoded relative to the previous, such that the receiver must only measure the difference in the arrival time of successive pulses. It is possible to limit the propagation of errors to adjacent symbols, so that an error in measuring the differential delay of one pulse will affect only two symbols, instead of affecting all successive measurements.

Sensitivity to Multipath Interference

Aside from the issues regarding receiver synchronization, the key disadvantage of PPM is that it is inherently sensitive to multipath interference that arises in channels with frequency-selective fading, whereby the receiver's signal contains one or more echoes of each transmitted pulse. Since the information is encoded in the time of arrival (either differentially, or relative to a common clock), the presence of one or more echoes can make it extremely difficult, if not impossible, to accurately determine the correct pulse position corresponding to the transmitted pulse. Multipath in Pulse Position Modulation systems can be easily mitigated by using the same techniques that are used in Radar systems that rely totally on synchronization and time of arrival of the received pulse to obtain their range position in the presence of echoes.

Non-coherent Detection

One of the principal advantages of PPM is that it is an M-ary modulation technique that can be implemented non-coherently, such that the receiver does not need to use a phase-locked loop (PLL) to track the phase of the carrier. This makes it a suitable candidate for optical communications systems, where coherent phase modulation and detection are difficult and extremely expensive. The only other common M-ary non-coherent modulation technique is M-ary frequency-shift keying (M-FSK), which is the frequency-domain dual to PPM.

PPM vs. M-FSK

PPM and M-FSK systems with the same bandwidth, average power, and transmission rate of M/T bits per second have identical performance in an additive white Gaussian noise (AWGN) channel. However, their performance differs greatly when comparing frequency-selective and frequency-flat fading channels. Whereas frequency-selective fading produces echoes that are highly disruptive for any of the M time-shifts used to encode PPM data, it selectively disrupts only some of the M possible frequency-shifts used to encode data for M-FSK. On the other hand, frequency-flat fading is more disruptive for M-FSK than PPM, as all M of the possible frequency-shifts are impaired by fading, while the short duration of the PPM pulse means that only a few of the M time-shifts are heavily impaired by fading.

Optical communications systems tend to have weak multipath distortions, and PPM is a viable modulation scheme in many such applications.

Applications for RF Communications

Narrowband RF (radio frequency) channels with low power and long wavelengths (i.e., low frequency) are affected primarily by flat fading, and PPM is better suited than M-FSK to be used in these scenarios. One common application with these channel characteristics, first used in the early 1960s with top-end HF (as low as 27 MHz) frequencies into the low-end VHF band frequencies (30 MHz to 75 MHz for RC use depending on location), is the radio control of model aircraft, boats and cars, originally known as "digital proportional" radio control. PPM is employed in these systems, with the position of each pulse representing the angular position of an analogue control on the transmitter, or possible states of a binary switch. The number of pulses per frame gives the number of controllable channels available. The advantage of using PPM for this type of application is that the electronics required to decode the signal are extremely simple, which leads to small, light-weight receiver/decoder units. (Model aircraft require parts that are as lightweight as possible). Servos made for model radio control include some of the electronics required to convert the pulse to the motor position – the receiver is required to first extract the information from the received radio signal through its intermediate frequency section, then demultiplex the separate channels from the serial stream, and feed the control pulses to each servo.

PPM Encoding for Radio Control

A complete PPM frame is about 22.5 ms (can vary between manufacturer), and signal low state is always 0.3 ms. It begins with a start frame (high state for more than 2 ms). Each channel (up to 8) is encoded by the time of the high state (PPM high state + 0.3 × (PPM low state) = servo PWM pulse width).

More sophisticated radio control systems are now often based on pulse-code modulation, which is more complex but offers greater flexibility and reliability. The advent of 2.4 GHz band FHSS radio-control systems in the early 21st century changed this still further.

Pulse-position modulation is also used for communication to the ISO/IEC 15693 contactless smart card, as well as the HF implementation of the Electronic Product Code (EPC) Class 1 protocol for RFID tags.

PULSE CODE MODULATION

Pulse code modulation is a method that is used to convert an analog signal into a digital signal, so that modified analog signal can be transmitted through the digital communication network. PCM is in binary form ,so there will be only two possible states high and low (0 and 1). We can also get back our analog signal by demodulation. The Pulse Code Modulation process is done in three steps Sampling, Quantization, and Coding. There are two specific types of pulse code modulations such as differential pulse code modulation (DPCM) and adaptive differential pulse code modulation (ADPCM).

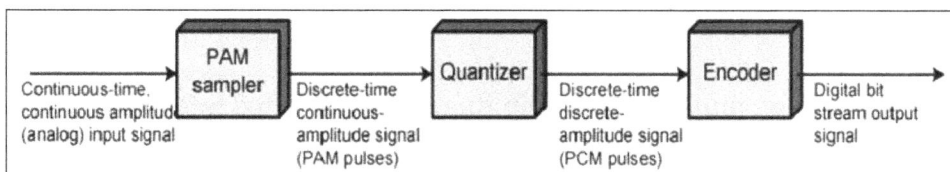

Block Diagram of PCM.

Here is a block diagram of the steps which are included in PCM. In sampling we are using PAM sampler that is Pulse Amplitude Modulation Sampler which converts continuous amplitude signal into Discrete-time- continuous signal (PAM pulses).

Pulse Code Modulation Theory

This above block diagram describes the whole process of PCM. The source of continuous time message signal is passed through a low pass filter and then sampling, Quantization, Encoding will be done.

Sampling

Sampling is a process of measuring the amplitude of a continuous-time signal at discrete instants, converts the continuous signal into a discrete signal. For example, conversion of a sound wave to a sequence of samples. The Sample is a value or set of values at a point in time or it can be spaced. Sampler extract samples of a continuous signal, it is a subsystem ideal sampler produces samples which are equivalent to the instantaneous value of the continuous signal at the specified various points. The Sampling process generates flat- top Pulse Amplitude Modulated (PAM) signal.

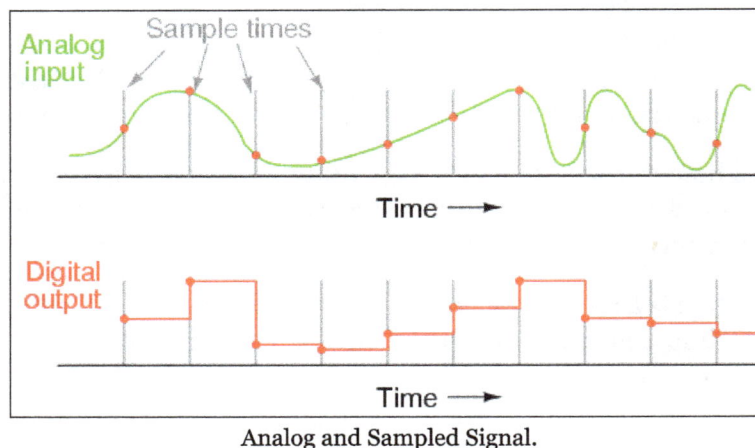

Analog and Sampled Signal.

Sampling frequency, Fs is the number of average samples per second also known as Sampling rate. According to the Nyquist Theorem sampling rate should be at least 2 times the upper cutoff frequency. Sampling frequency, Fs>=2*fmax to avoid Aliasing Effect. If the sampling frequency is very higher than the Nyquist rate it become Oversampling, theoretically a bandwidth limited signal can be reconstructed if sampled at above the Nyquist rate. If the sampling frequency is less than the Nyquist rate it will become Undersampling.

Basically two types of techniques are used for the sampling process. Those are: 1) Natural Sampling and 2) Flat- top Sampling.

Quantization

In quantization, an analog sample with an amplitude that converted into a digital sample with an amplitude that takes one of a specific defined set of quantization values. Quantization is done by dividing the range of possible values of the analog samples into some different levels, and assigning the center value of each level to any sample in quantization interval. Quantization approximates the analog sample values with the nearest quantization values. So almost all the quantized samples will differ from the original samples by a small amount. That amount is called as quantization error. The result of this quantization error is we will hear hissing noise when play a random signal. Converting analog samples into binary numbers that is 0 and 1.

In most of the cases we will use uniform quantizers. Uniform quantization is applicable when the sample values are in a finite range (Fmin, Fmax). The total data range is divided into 2n levels, let it be L intervals. They will have an equal length Q. Q is known as Quantization interval or quantization step size. In uniform quantization there will be no quantization error.

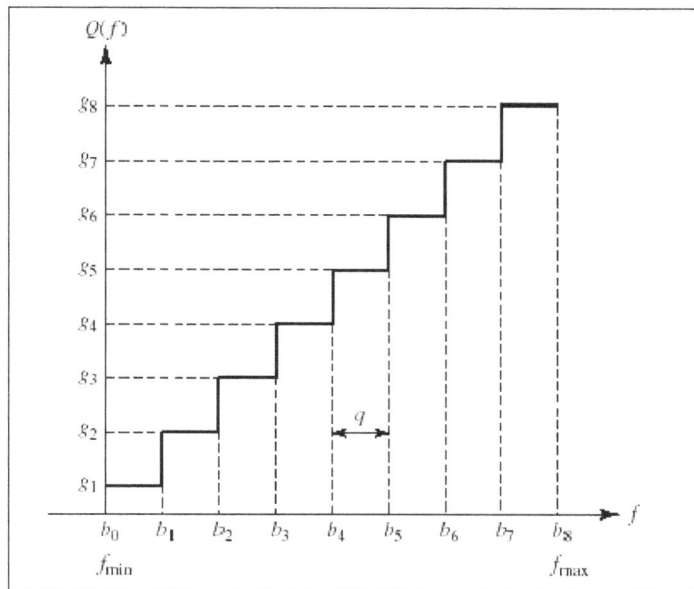

Uniformly Quantized Signal.

As we know,

L=2n, then Step size $Q = (Fmax - Fmin) / L$

Interval i is mapped to the middle value. We will store or send only index value of quantized value.

An Index value of quantized value $Qi (F) = [F - Fmin / Q]$.

Quantized value $Q (F) = Qi (F) Q + Q / 2 + Fmin$.

But there are some problems raised in uniform quantization those are:

- Only optimal for uniformly distributed signal.

- Real audio signals are more concentrated near zeros.

- The Human ear is more sensitive to quantization errors at small values.

The solution for this problem is using Non- uniform quantization. In this Process quantization interval is smaller near zero.

Coding

The encoder encodes the quantized samples. Each quantized sample is encoded into an 8-bit code word by using A-law in the encoding process.

- Bit 1 is the most significant bit (MSB), it represents the polarity of the sample. "1" represents positive polarity and "0" represents negative polarity.

- Bit 2, 3 and 4 will defines the location of sample value. These three bits together form linear curve for low level negative or positive samples.

- Bit 5, 6, 7 and 8 are the least significant bits (LSB) it represents one of the segments quantized value. Each segment is divided into 16 quantum levels.

PCM is two types Differential Pulse Code Modulation (DPCM) and Adaptive Differential Pulse Code Modulation (ADPCM). In DPCM only the difference between a sample and the previous value is encoded. The difference will be much smaller than the total sample value so we need some bits for getting same accuracy as in ordinary PCM. So that the required bit rate will also reduce. For example, in 5 bit code 1 bit is for polarity and the remaining 4 bits for 16 quantum levels.

ADPCM is achieved by adapting the quantizing levels to analog signal characteristics. We can estimate the values with preceding sample values. Error estimation is done as same as in DPCM. In 32Kbps ADPCM method difference between predicted value and sample value is coded with 4 bits, so that we'll get 15 quantum levels. In this method data rate is half of the conventional PCM.

Pulse Code Demodulation

Pulse Code Demodulation will be doing the same modulation process in reverse. Demodulation starts with decoding process, during transmission the PCM signal will effected by the noise interference. So, before the PCM signal sends into the PCM demodulator, we have to recover the signal into the original level for that we are using a comparator. The PCM signal is a series pulse wave signal, but for demodulation we need wave to be parallel.

By using a serial to parallel converter the series pulse wave signal will be converted into a parallel digital signal. After that the signal will pass through n-bits decoder, it should be a Digital to Analog converter. Decoder recovers the original quantization values of the digital signal. This quantization value also includes a lot of high frequency harmonics with original audio signals. For avoiding unnecessary signals we utilize a low-pass filter at the final part.

Pulse Code Modulation Advantages

- Analog signal can be transmitted over a high- speed digital communication system.

- Probability of occurring error will reduce by the use of appropriate coding methods.

- PCM is used in Telkom system, digital audio recording, digitized video special effects, digital video, voice mail.

- PCM is also used in Radio control units as transmitter and also receiver for remote controlled cars, boats, planes.

- The PCM signal is more resistant to interference than normal signal.

Pulse Code Modulation Technique

Pulse code Modulation: The pulse code modulator technique samples the input signal x(t) at a sampling frequency. This sampled variable amplitude pulse is then digitalized by the analog to digital Converter shows the PCM generator.

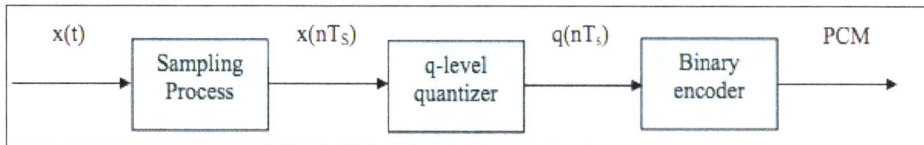

In the PCM generator, the signal is first passed through sampler which is sampled at a rate of (f_s) where:

$$f_s \geq 2f_m$$

The output of the sampler $x(nT_s)$ which is discrete in time is fed to a qlevel quantizer. The quantizer compares the input $x(nT_s)$ with it's fixed levels. It assigns any one of the digital level to $x(nT_s)$ that results in minimum distortion or error. The error is called quantization error, thus the output of the quantizer is a digital level called $q(nT_s)$. The quantized signal level q(nTs) is binary encode. The encoder converts the input signal to v digits binary word.

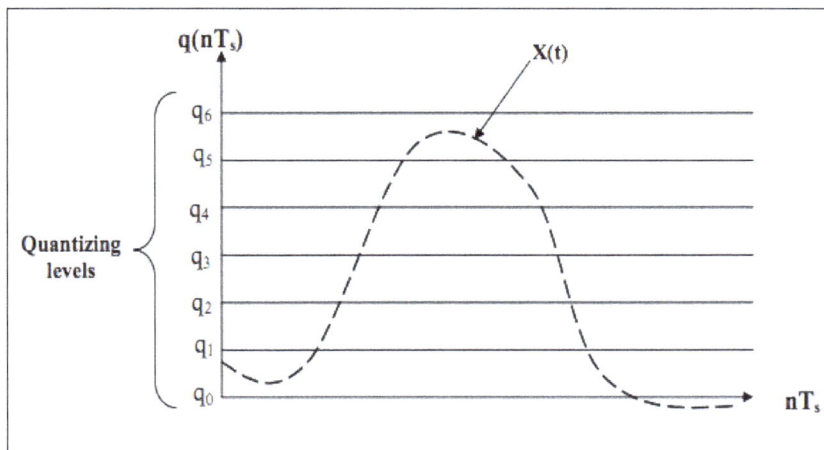

A sampled signal and the quantized levels.

Figure below shows the block diagram of the PCM receiver. The receiver starts by reshaping the received pulses, removes the noise and then converts the binary bits to analog. The received samples are then filtered by a low pass filter; the cut off frequency is at f_c,

$$f_c = f_m$$

where f_m is the highest frequency component in the original signal.

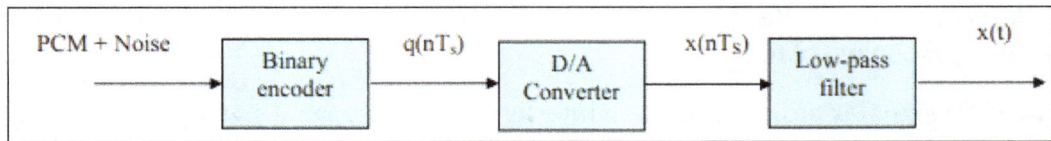

PCM demodulator.

It is impossible to reconstruct the original signal x(t) because of the permanent quantization error introduced during quantization at the transmitter. The quantization error can be reduced by the increasing quantization levels. This corresponds to the increase of bits per sample. But increasing bits (v) increases the signaling rate and requires a large transmission bandwidth. The choice of the parameter for the number of quantization levels must be acceptable with the quantization noise (quantization error) shows the reconstructed signal.

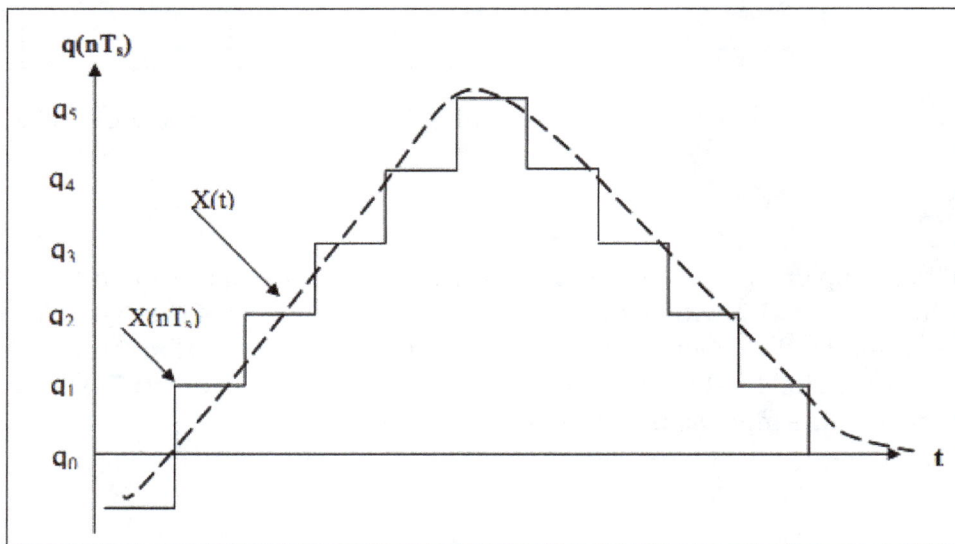

The reconstructed signal.

Signaling Rate in PCM

Let the quantizer use 'v' number of binary digits to represent each level. Then the number of levels that can be represented by v digits will be:

$$q = 2^v$$

The number of bits per second is given by:

(Number of bits per second) = (Number of bits per samples) × (number of samples per second) = v (bits per sample) x fs (samples per second) The number of bits per second is also called signaling rate of PCM and is denoted by 'r",

Signaling rate= $v\, f_s$

where,

$f_s \geq f_m$

Example:

If the number of binary bits = 3 and the sampling rate is 2 sample/sec, find the signaling rate, number of quantization levels?

Solution:

$$f_s = 2, \ v = 3$$

signaling rate $(r) = v \ f_s$

$$= 3 \times 2$$

$$= 6 \text{ bits/sec}$$

Number of quantization $(q) = 2^v$,

$$= 2^3$$

$$= 8 \text{ levels}$$

Quantization Noise in PCM System

Errors are introduced in the signal because of the quantization process. This error is called "quantization error". We define the quantization error as:

$$\varepsilon = x_q \ (nT_s) - x(nT_s)$$

Let an input signal $x(nT_s)$ have an amplitude in the range of x_{max} to $- x_{max}$.

The total amplitude range is:

Total amplitude $= x_{max} - (-x_{max})$

$$= 2 \ x_{max}$$

If the amplitude range is divided into 'q' levels of quantizer, then the step size 'Δ'.

$$\Delta = \frac{2X_{max}}{q}$$

If the minimum and maximum values are equal to 1, $x_{max} = 1$, $- x_{max} = -1$,

then the equation will be:

$$\Delta = \frac{2}{q}$$

If Δ is small it can be assumed that the quantization error is uniformly distributed. The quantization noise is uniformly distributed in the interval $[-\Delta/2, \Delta/2]$. The figure shows the uniform distribution of quantization noise.

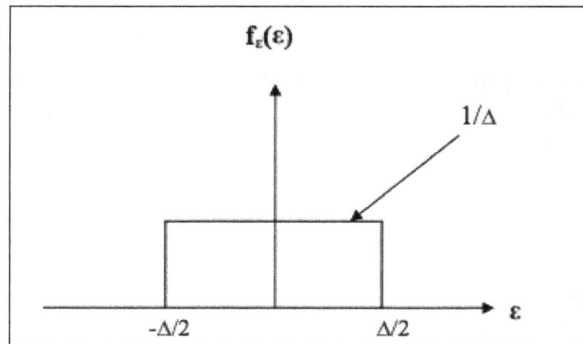

The noise power is given by:

$$\text{Noise power} = V^2_{noise} / R$$

V^2_{noise} : is the mean square value of noise voltage, since noise is defined by random variable "ε" and PDF $f\varepsilon(\varepsilon)$, it's mean square value is given by,

$$V^2_{noise} = \int_{-\Delta/2}^{\Delta/2} \varepsilon^2 f_e(\varepsilon).d\varepsilon$$

Substitute the value of $f\varepsilon(\varepsilon) = 1/\Delta$ in above equation,

$$V^2_{noise} = \int_{-\Delta/2}^{\Delta/2} \varepsilon^2 1/\Delta.d\varepsilon$$

$$= \frac{1}{3\Delta}\left[\frac{\Delta^3}{8} + \frac{\Delta^3}{8}\right]$$

$$= \frac{\Delta^2}{12}$$

If R = 1,

$$\text{Quantization noise power} = \frac{\Delta^2}{12}$$

Signal to Quantization Noise Ratio in PCM

The signal to quantization noise ratio is given as:

$$\frac{S}{N_q} = \frac{\text{Normalized signal power}}{\text{Normalized noise power}}$$

$$= \frac{\text{Normalized signal power}}{\dfrac{\Delta^2}{12}}$$

The number of quantization value is equal to:

$$q = 2^v$$

Putting this value in equation $\Delta = \dfrac{2X_{max}}{q}$, we get,

$$\Delta = \dfrac{2X \max}{2^v}$$

Substitute this value in equation $\dfrac{\text{Normalized signal power}}{\dfrac{\Delta^2}{12}}$, we get,

$$\frac{S}{N_q} = \frac{\text{Normalized signal power}}{\left[\dfrac{2X_{max}}{2^v}\right] * \dfrac{1}{12}}$$

Let the normalized signal power is equal to P then the signal to quantization noise will be given by:

$$\frac{S}{N_q} = \frac{3P * 2^{2v}}{X^2_{max}}$$

Examples:

A signal that has the highest frequency component of 4.2 MHz and a peak to peak value of 4 volts is transmitted using a binary PCM. The number of quantization levels is 512 and P = 0.04 W calculate:

1. Code word length.

2. Bite rate.

3. Output signal to quantization noise ratio.

Solution:

$$f_m = 4.2 \text{ MHz, q=512}$$

$$512 = 2^v$$

1. length of code word= v =9 bits,

2. Bit rate $r = v f_s$

$$= v * (2 f_m)$$

$$= 9 \times 2 \times 4.2 \text{ MHz}$$

$$= 75.6 \times 10^6 \text{ bits/sec}$$

$$\frac{s}{N_q} = \frac{3p \times 2^2 v}{x^2_{max}}$$

$$= \frac{3 \times 0.04 \times 2^{16}}{4}$$

$$= 1966.08 \approx 33\text{dB}$$

Advantages of PCM

1. Effect of noise is reduced.

2. PCM permits the use of pulse regeneration.

3. Multiplexing of various PCM signals is possible.

DIFFERENTIAL PULSE CODE MODULATION

Differential pulse code modulation is a technique of analog to digital signal conversion. This technique samples the analog signal and then quantizes the difference between the sampled value and its predicted value, then encodes the signal to form a digital value. Before going to discuss differential pulse code modulation, we have to know the demerits of PCM (Pulse Code Modulation). The samples of a signal are highly correlated with each other. The signal's value from the present sample to next sample does not differ by a large amount. The adjacent samples of the signal carry the same information with a small difference. When these samples are encoded by the standard PCM system, the resulting encoded signal contains some redundant information bits. The below figure illustrates this.

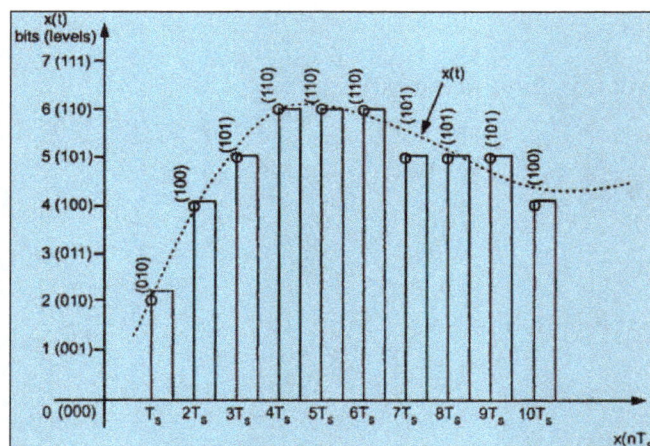

Redundant information Bits in PCM

The above figure shows a continuing time signal x(t) denoted by a dotted line. This signal is sampled by flat top sampling at intervals Ts, 2Ts, 3Ts...nTs. The sampling frequency is selected to be higher than the Nyquist rate. These samples are encoded by using 3-bit (7 levels) PCM. The samples which are quantized to nearest digital level as shown by small circles in the above figure. The encoded binary value of each sample is written on the top of the samples. Just observe the above figure at samples taken at 4Ts, 5Ts, and 6Ts are encoded to the same value of (110). This information can be carried only by one sample value. But three samples are carrying the same information means redundant. Now let consider the samples at 9Ts and 10Ts, the difference between these samples only due to the last bit and first two bits are redundant since they do not change. So in order to make the process this redundant information and to have a better output. It is an intelligent decision to take a predicted sampled value, assumed from its previous output and summarise them with the quantized values. Such a process is called as Differential PCM (DPCM) technique.

Principle of Differential Pulse Code Modulation

If the redundancy is reduced, then the overall bitrate will decrease and the number of bits required to transmit one sample will also reduce. This type of digital pulse modulation technique is called differential pulse code modulation. The DPCM works on the principle of prediction. The value of the present sample is predicted from the previous samples. The prediction may not be exact, but it is very close to the actual sample value.

Differential Pulse Code Modulation Transmitter

The below figure shows the DPCM transmitter. The transmitter consists of a comparator, quantizer, prediction filter and an encoder.

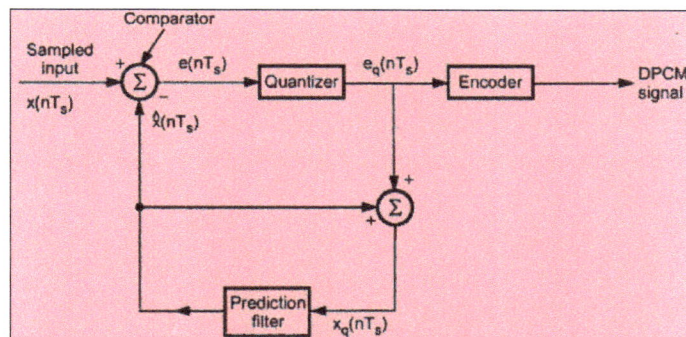

Differential Pulse Code Modulator.

The sampled signal is denoted by $x(nTs)$ and the predicted signal is indicated by $\hat{x}(nTs)$ The comparator finds out the difference between the actual sample value $x(nTs)$ and the predicted value $\hat{x}(nTs)$. This is called signal error and it is denoted as $e(nTs)$,

$$e(nTs) = x(nTs) - \hat{x}(nTs)$$

Here the predicted value $\hat{x}(nTs)$ is produced by using a prediction filter (signal processing filter). The quantizer output signal $eq(nTs)$ and the previous prediction is added and given as input to the prediction filter, this signal is denoted by $xq(nTs)$ This makes the prediction closer to the actually sampled signal. The quantized error signal $eq(nTs)$ is very small and can be encoded by using a small number of bits. Thus the number of bits per sample is reduced in DPCM.

The quantizer output would be written as,

$$eq(nTs) = e(nTs) + q(nTs)$$

Here $q(nTs)$ is quantization error. From the above block diagram the prediction filter input $xq(nTs)$ is obtained by sum of $x^{\wedge}(nTs)$ and the quantizer output $eq(nTs)$.

i.e, $xq(nTs) = \hat{x}(nTs)+eq(nTs)$

by substituting the value of $eq(nTs)$ from the $eq(nTs)= e(nTs)+ q(nTs)$ in $xq(nTs) = \hat{x}(nTs)+ eq(nTs)$ we get,

$xq(nTs) = \hat{x}(nTs)+e(nTs)+q(nTs)$

Equation $e(nTs) = x(nTs)- \hat{x}(nTs)$ can written as,

$e(nTs)=x^{\wedge}(nTs)=x(nTs)$

from the above equations $xq(nTs) = \hat{x}(nTs)+e(nTs)+q(nTs)$ and $e(nTs)=x^{\wedge}(nTs)=x(nTs)$ we get,

$xq(nTs) = x(nTs)+ x(nTs)$

Therefore, the quantized version of signal $xq(nTs)$ is the sum of original sample value and quantized error $q(nTs)$. The quantized error can be positive or negative. So the output of the prediction filter does not depend on its characteristics.

Differential Pulse Code Modulation Receiver

In order to reconstruct the received digital signal, the DPCM receiver consists of a decoder and prediction filter. In the absenteeism of noise, the encoded receiver input will be the same as the encoded transmitter output.

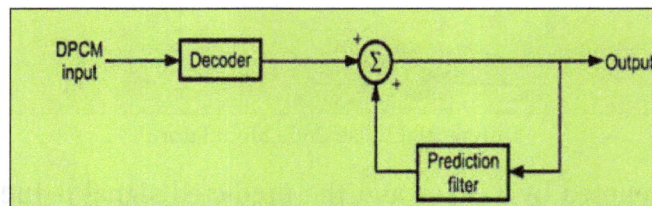

Differential Pulse Code Modulation Receiver.

As we discussed above, the predictor undertakes a value, based on the previous outputs. The input given to the decoder is processed and that output is summed up with the output of the predictor, to obtain a better output. That means here first of all the decoder will reconstruct the quantized form of original signal. Therefore the signal at the receiver differs from the actual signal by quantization error $q(nTs)$, which is introduced permanently in the reconstructed signal.

S. NO	Parameters	Pulse Code Modulation (PCM)	Differential Pulse Code Modulation (DPCM)
1	Number of bits	It uses 4, 8, or 16 bits per sample	< PCM bits
2	Levels, step size	Fixed step size. Cannot varied	A fixed number of levels are used.
3	Bit redundancy	Present	Can permanently remove

4	Quantization error and distortion	Depends on the number of levels used	Slope overload distortion and quantization noise are present, but very less as compared to PCM
5	Bandwidth of transmission channel	Higher bandwidth has been required since number of bits are absent	Lower than PCM bandwidth
6	Feedback	No feedback in Tx and Rx	Feedback exists
7	Complexity of notation	Complex	Simple
8	Signal to noise ratio (SNR)	Good	Fair

Applications of DPCM

The DPCM technique mainly used Speech, image and audio signal compression. The DPCM conducted on signals with the correlation between successive samples leads to good compression ratios. In images, there is a correlation between the neighbouring pixels, in video signals, the correlation is between the same pixels in consecutive frames and inside frames (which is same as correlation inside the image).

This method is suitable for real Time applications. To understand the efficiency of this method of medical compression and real-time application of medical imaging such as telemedicine and online diagnosis. Therefore, it can be efficient for lossless compression and implementation for lossless or near-lossless medical image compression.

DELTA MODULATION

A delta modulation (DM or Δ-modulation) is an analog-to-digital and digital-to-analog signal conversion technique used for transmission of voice information where quality is not of primary importance. DM is the simplest form of differential pulse-code modulation (DPCM) where the difference between successive samples are encoded into n-bit data streams. In delta modulation, the transmitted data are reduced to a 1-bit data stream. Its main features are:

- The analog signal is approximated with a series of segments.

- Each segment of the approximated signal is compared to the preceding bits and the successive bits are determined by this comparison.

- Only the change of information is sent, that is, only an increase or decrease of the signal amplitude from the previous sample is sent whereas a no-change condition causes the modulated signal to remain at the same 0 or 1 state of the previous sample.

To achieve high signal-to-noise ratio, delta modulation must use oversampling techniques, that is, the analog signal is sampled at a rate several times higher than the Nyquist rate.

Derived forms of delta modulation are continuously variable slope delta modulation, delta-sigma modulation, and differential modulation. Differential pulse-code modulation is the superset of DM.

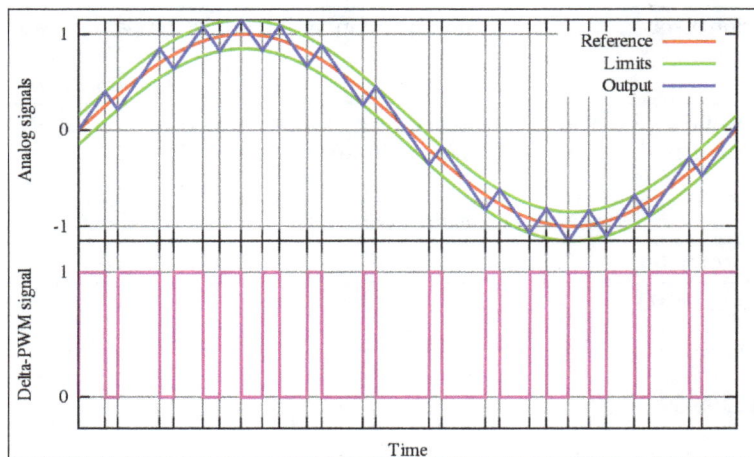

Principle of the delta PWM. The output signal (blue) is compared with the limits (green). The limits (green) correspond to the reference signal (red), offset by a given value. Every time the output signal reaches one of the limits, the PWM signal changes state.

Principle

Rather than quantizing the absolute value of the input analog waveform, delta modulation quantizes the difference between the current and the previous step, as shown in the block diagram in figure.

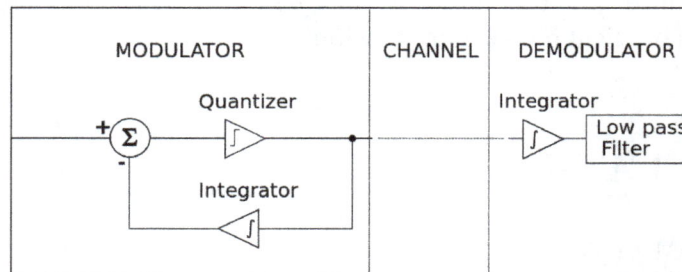

Block diagram of a Δ-modulator/demodulator.

The modulator is made by a quantizer which converts the difference between the input signal and the average of the previous steps. In its simplest form, the quantizer can be realized with a comparator referenced to 0 (two levels quantizer), whose output is 1 or 0 if the input signal is positive or negative. It is also a bit-quantizer as it quantizes only a bit at a time. The demodulator is simply an integrator (like the one in the feedback loop) whose output rises or falls with each 1 or 0 received. The integrator itself constitutes a low-pass filter.

Transfer Characteristics

The transfer characteristics of a delta modulated system follows a signum function, as it quantizes only two levels and also one-bit at a time.

The two sources of noise in delta modulation are "slope overload", when step size is too small to track the original waveform, and "granularity", when step size is too large. But a 1971 study shows that slope overload is less objectionable compared to granularity than one might expect based solely on SNR measures.

Output Signal Power

In delta modulation there is a restriction on the amplitude of the input signal, because if the transmitted signal has a large derivative (abrupt changes) then the modulated signal can not follow the input signal and slope overload occurs. E.g. if the input signal is,

$$m(t) = A\cos(\omega t),$$

the modulated signal (derivative of the input signal) which is transmitted by the modulator is,

$$|\dot{m}(t)|_{max} = \omega A,$$

whereas the condition to avoid slope overload is,

$$|\dot{m}(t)|_{max} = \omega A < \sigma f_s.$$

So the maximum amplitude of the input signal can be,

$$A_{max} = \frac{\sigma f_s}{\omega},$$

where f_s is the sampling frequency and ω is the frequency of the input signal and σ is step size in quantization. So A_{max} is the maximum amplitude that DM can transmit without causing the slope overload and the power of transmitted signal depends on the maximum amplitude.

Bit-rate

If the communication channel is of limited bandwidth, there is the possibility of interference in either DM or PCM. Hence, 'DM' and 'PCM' operate at same bit-rate which is equal to N times the sampling frequency.

Adaptive Delta Modulation

Adaptive delta modulation (ADM) was first published by Dr. John E. Abate (AT&T Bell Laboratories Fellow) in his doctoral thesis at NJ Institute Of Technology in 1968. ADM was later selected as the standard for all NASA communications between mission control and space-craft.

Adaptive delta modulation or Continuously variable slope delta modulation (CVSD) is a modification of DM in which the step size is not fixed. Rather, when several consecutive bits have the same direction value, the encoder and decoder assume that slope overload is occurring, and the step size becomes progressively larger.

Otherwise, the step size becomes gradually smaller over time. ADM reduces slope error, at the expense of increasing quantizing error. This error can be reduced by using a low-pass filter. ADM provides robust performance in the presence of bit errors meaning error detection and correction are not typically used in an ADM radio design, it is this very useful technique that allows for adaptive-delta-modulation.

Applications

Contemporary applications of Delta Modulation includes, but is not limited to, recreating legacy synthesizer waveforms. With the increasing availability of FPGAs and game-related ASICs, sample rates are easily controlled so as to avoid slope overload and granularity issues. For example, the C64DTV used a 32 MHz sample rate, providing ample dynamic range to recreate the SID output to acceptable levels.

SBS Application 24 kbps Delta Modulation

Delta Modulation was used by Satellite Business Systems or SBS for its voice ports to provide long distance phone service to large domestic corporations with a significant inter-corporation communications need (such as IBM). This system was in service throughout the 1980s. The voice ports used digitally implemented 24 kbit/s delta modulation with Voice Activity Compression (VAC) and echo suppressors to control the half second echo path through the satellite. They performed formal listening tests to verify the 24 kbit/s delta modulator achieved full voice quality with no discernible degradation as compared to a high quality phone line or the standard 64 kbit/s μ-law companded PCM. This provided an eight to three improvement in satellite channel capacity. IBM developed the Satellite Communications Controller and the voice port functions.

The original proposal in 1974, used a state-of-the-art 24 kbit/s delta modulator with a single integrator and a Shindler Compander modified for gain error recovery. This proved to have less than full phone line speech quality. In 1977, one engineer with two assistants in the IBM Research Triangle Park, NC laboratory was assigned to improve the quality.

The final implementation replaced the integrator with a Predictor implemented with a two pole complex pair low-pass filter designed to approximate the long term average speech spectrum. The theory was that ideally the integrator should be a predictor designed to match the signal spectrum. A nearly perfect Shindler Compander replaced the modified version. It was found the modified compander resulted in a less than perfect step size at most signal levels and the fast gain error recovery increased the noise as determined by actual listening tests as compared to simple signal to noise measurements. The final compander achieved a very mild gain error recovery due to the natural truncation rounding error caused by twelve bit arithmetic.

The complete function of delta modulation, VAC and Echo Control for six ports was implemented in a single digital integrated circuit chip with twelve bit arithmetic. A single digital-to-analog converter (DAC) was shared by all six ports providing voltage compare functions for the modulators and feeding sample and hold circuits for the demodulator outputs. A single card held the chip, DAC and all the analog circuits for the phone line interface including transformers.

TIME DIVISION MULTIPLEXING

Time-division multiplexing (TDM) is a method of transmitting and receiving independent signals over a common signal path by means of synchronized switches at each end of the transmission line so that each signal appears on the line only a fraction of time in an alternating pattern. It is used when the bit rate of the transmission medium exceeds that of the signal to be transmitted. This form of signal multiplexing was developed in telecommunications for telegraphy systems in the late 19th century, but found its most common application in digital telephony in the second half of the 20th century.

Technology

Time-division multiplexing is used primarily for digital signals, but may be applied in analog multiplexing in which two or more signals or bit streams are transferred appearing simultaneously as sub-channels in one communication channel, but are physically taking turns on the channel. The time domain is divided into several recurrent *time slots* of fixed length, one for each sub-channel. A sample byte or data block of sub-channel 1 is transmitted during time slot 1, sub-channel 2 during time slot 2, etc. One TDM frame consists of one time slot per sub-channel plus a synchronization channel and sometimes error correction channel before the synchronization. After the last sub-channel, error correction, and synchronization, the cycle starts all over again with a new frame, starting with the second sample, byte or data block from sub-channel 1, etc.

Application Examples

- The plesiochronous digital hierarchy (PDH) system, also known as the PCM system, for digital transmission of several telephone calls over the same four-wire copper cable (T-carrier or E-carrier) or fiber cable in the circuit switched digital telephone network.

- The synchronous digital hierarchy (SDH)/synchronous optical networking (SONET) network transmission standards that have replaced PDH.

- The Basic Rate Interface and Primary Rate Interface for the Integrated Services Digital Network (ISDN).

- The RIFF (WAV) audio standard interleaves left and right stereo signals on a per-sample basis.

TDM can be further extended into the time-division multiple access (TDMA) scheme, where several stations connected to the same physical medium, for example sharing the same frequency channel, can communicate. Application examples include:

- The GSM telephone system.

- The Tactical Data Links Link 16 and Link 22.

Multiplexed Digital Transmission

In circuit-switched networks, such as the public switched telephone network (PSTN), it is desirable to transmit multiple subscriber calls over the same transmission medium to effectively utilize the bandwidth of the medium. TDM allows transmitting and receiving telephone switches to create channels (*tributaries*) within a transmission stream. A standard DS0 voice signal has a data bit rate of 64 kbit/s. A TDM circuit runs at a much higher signal bandwidth, permitting the bandwidth to be divided into time frames (time slots) for each voice signal which is multiplexed onto the line by the transmitter. If the TDM frame consists of n voice frames, the line bandwidth is $n*64$ kbit/s.

Each voice time slot in the TDM frame is called a channel. In European systems, standard TDM frames contain 30 digital voice channels (E1), and in American systems (T1), they contain 24 channels. Both standards also contain extra bits (or bit time slots) for signaling and synchronization bits.

Multiplexing more than 24 or 30 digital voice channels is called *higher order multiplexing*. Higher order multiplexing is accomplished by multiplexing the standard TDM frames. For example, a European 120 channel TDM frame is formed by multiplexing four standard 30 channel TDM frames. At each higher order multiplex, four TDM frames from the immediate lower order are combined, creating multiplexes with a bandwidth of $n*64$ kbit/s, where n = 120, 480, 1920, etc.

Telecommunications Systems

There are three types of synchronous TDM: T1, SONET/SDH, and ISDN.

Plesiochronous digital hierarchy (PDH) was developed as a standard for multiplexing higher order frames. PDH created larger numbers of channels by multiplexing the standard Europeans 30 channel TDM frames. This solution worked for a while; however PDH suffered from several inherent drawbacks which ultimately resulted in the development of the Synchronous Digital Hierarchy (SDH). The requirements which drove the development of SDH were these:

- Be synchronous – All clocks in the system must align with a reference clock.

- Be service-oriented – SDH must route traffic from End Exchange to End Exchange without worrying about exchanges in between, where the bandwidth can be reserved at a fixed level for a fixed period of time.

- Allow frames of any size to be removed or inserted into an SDH frame of any size.

- Easily manageable with the capability of transferring management data across links.

- Provide high levels of recovery from faults.

- Provide high data rates by multiplexing any size frame, limited only by technology.

- Give reduced bit rate errors.

SDH has become the primary transmission protocol in most PSTN networks. It was developed to allow streams 1.544 Mbit/s and above to be multiplexed, in order to create larger SDH frames known as Synchronous Transport Modules (STM). The STM-1 frame consists of smaller streams that are multiplexed to create a 155.52 Mbit/s frame. SDH can also multiplex packet based frames e.g. Ethernet, PPP and ATM.

While SDH is considered to be a transmission protocol (Layer 1 in the OSI Reference Model), it also performs some switching functions, as stated in the third bullet point requirement listed above. The most common SDH Networking functions are these:

- SDH Crossconnect – The SDH Crossconnect is the SDH version of a Time-Space-Time crosspoint switch. It connects any channel on any of its inputs to any channel on any of its outputs. The SDH Crossconnect is used in Transit Exchanges, where all inputs and outputs are connected to other exchanges.

- SDH Add-Drop Multiplexer – The SDH Add-Drop Multiplexer (ADM) can add or remove any multiplexed frame down to 1.544Mb. Below this level, standard TDM can be performed. SDH ADMs can also perform the task of an SDH Crossconnect and are used in End Exchanges where the channels from subscribers are connected to the core PSTN network.

SDH network functions are connected using high-speed optic fibre. Optic fibre uses light pulses to transmit data and is therefore extremely fast. Modern optic fibre transmission makes use of wavelength-division multiplexing (WDM) where signals transmitted across the fibre are transmitted at different wavelengths, creating additional channels for transmission. This increases the speed and capacity of the link, which in turn reduces both unit and total costs.

Statistical Time-division Multiplexing

Statistical time-division multiplexing (STDM) is an advanced version of TDM in which both the address of the terminal and the data itself are transmitted together for better routing. Using STDM allows bandwidth to be split over one line. Many college and corporate campuses use this type of TDM to distribute bandwidth.

On a 10-Mbit line entering a network, STDM can be used to provide 178 terminals with a dedicated 56k connection (178 * 56k = 9.96Mb). A more common use however is to only grant the bandwidth when that much is needed. STDM does not reserve a time slot for each terminal, rather it assigns a slot when the terminal is requiring data to be sent or received.

In its primary form, TDM is used for circuit mode communication with a fixed number of channels and constant bandwidth per channel. Bandwidth reservation distinguishes time-division multiplexing from statistical multiplexing such as statistical time-division multiplexing. In pure TDM, the time slots are recurrent in a fixed order and pre-allocated to the channels, rather than scheduled on a packet-by-packet basis.

In dynamic TDMA, a scheduling algorithm dynamically reserves a variable number of time slots in each frame to variable bit-rate data streams, based on the traffic demand of each data stream. Dynamic TDMA is used in:

- HIPERLAN/2,
- Dynamic synchronous transfer mode,
- IEEE 802.16a.

Asynchronous time-division multiplexing (ATDM), is an alternative nomenclature in which STDM designates synchronous time-division multiplexing, the older method that uses fixed time slots.

FREQUENCY DIVISION MULTIPLEXING

In telecommunications, frequency-division multiplexing (FDM) is a technique by which the total bandwidth available in a communication medium is divided into a series of non-overlapping frequency bands, each of which is used to carry a separate signal. This allows a single transmission medium such as a cable or optical fiber to be shared by multiple independent signals. Another use is to carry separate serial bits or segments of a higher rate signal in parallel.

The most natural example of frequency-division multiplexing is radio and television broadcasting, in which multiple radio signals at different frequencies pass through the air at the same time.

Another example is cable television, in which many television channels are carried simultaneously on a single cable. FDM is also used by telephone systems to transmit multiple telephone calls through high capacity trunklines, communications satellites to transmit multiple channels of data on uplink and downlink radio beams, and broadband DSL modems to transmit large amounts of computer data through twisted pair telephone lines, among many other uses.

An analogous technique called wavelength division multiplexing is used in fiber-optic communication, in which multiple channels of data are transmitted over a single optical fiber using different wavelengths (frequencies) of light.

Working

The multiple separate information (modulation) signals that are sent over an FDM system, such as the video signals of the television channels that are sent over a cable TV system, are called baseband signals. At the source end, for each frequency channel, an electronic oscillator generates a *carrier* signal, a steady oscillating waveform at a single frequency that serves to "carry" information. The carrier is much higher in frequency than the baseband signal. The carrier signal and the baseband signal are combined in a modulator circuit. The modulator alters some aspect of the carrier signal, such as its amplitude, frequency, or phase, with the baseband signal, "piggybacking" the data onto the carrier.

The result of modulating (mixing) the carrier with the baseband signal is to generate sub-frequencies near the carrier frequency, at the sum $(f_C + f_B)$ and difference $(f_C - f_B)$ of the frequencies. The information from the modulated signal is carried in sidebands on each side of the carrier frequency. Therefore, all the information carried by the channel is in a narrow band of frequencies clustered around the carrier frequency, this is called the passband of the channel.

Similarly, additional baseband signals are used to modulate carriers at other frequencies, creating other channels of information. The carriers are spaced far enough apart in frequency that the band of frequencies occupied by each channel, the passbands of the separate channels, do not overlap. All the channels are sent through the transmission medium, such as a coaxial cable, optical fiber, or through the air using a radio transmitter. As long as the channel frequencies are spaced far enough apart that none of the passbands overlap, the separate channels will not interfere with each other. Thus the available bandwidth is divided into "slots" or channels, each of which can carry a separate modulated signal.

For example, the coaxial cable used by cable television systems has a bandwidth of about 1000 MHz, but the passband of each television channel is only 6 MHz wide, so there is room for many channels on the cable (in modern digital cable systems each channel in turn is subdivided into subchannels and can carry up to 10 digital television channels).

At the destination end of the cable or fiber, or the radio receiver, for each channel a local oscillator produces a signal at the carrier frequency of that channel, that is mixed with the incoming modulated signal. The frequencies subtract, producing the baseband signal for that channel again. This is called demodulation. The resulting baseband signal is filtered out of the other frequencies and output to the user.

Telephone

For long distance telephone connections, 20th century telephone companies used L-carrier and similar coaxial cable systems carrying thousands of voice circuits multiplexed in multiple stages by channel banks.

For shorter distances, cheaper balanced pair cables were used for various systems including Bell System K- and N-Carrier. Those cables didn't allow such large bandwidths, so only 12 voice channels (double sideband) and later 24 (single sideband) were multiplexed into four wires, one pair for each direction with repeaters every several miles, approximately 10 km. By the end of the 20th Century, FDM voice circuits had become rare. Modern telephone systems employ digital transmission, in which time-division multiplexing (TDM) is used instead of FDM.

Since the late 20th century digital subscriber lines (DSL) have used a Discrete multitone (DMT) system to divide their spectrum into frequency channels.

The concept corresponding to frequency-division multiplexing in the optical domain is known as wavelength-division multiplexing.

Group and Supergroup

A once commonplace FDM system, used for example in L-carrier, uses crystal filters which operate at the 8 MHz range to form a Channel Group of 12 channels, 48 kHz bandwidth in the range 8140 to 8188 kHz by selecting carriers in the range 8140 to 8184 kHz selecting upper sideband this group can then be translated to the standard range 60 to 108 kHz by a carrier of 8248 kHz. Such systems are used in DTL (Direct To Line) and DFSG (Directly formed super group).

132 voice channels (2SG + 1G) can be formed using DTL plane the modulation and frequency plan are given in FIG1 and FIG2 use of DTL technique allows the formation of a maximum of 132 voice channels that can be placed direct to line. DTL eliminates group and super group equipment.

DFSG can take similar steps where a direct formation of a number of super groups can be obtained in the 8 kHz the DFSG also eliminates group equipment and can offer:

- Reduction in cost 7% to 13%.

- Less equipment to install and maintain.

- Increased reliability due to less equipment.

Both DTL and DFSG can fit the requirement of low density system (using DTL) and higher density system (using DFSG). The DFSG terminal is similar to DTL terminal except instead of two super groups many super groups are combined. A Mastergroup of 600 channels (10 super-groups) is an example based on DFSG.

Other Examples:

FDM can also be used to combine signals before final modulation onto a carrier wave. In this case the carrier signals are referred to as subcarriers: an example is stereo FM transmission, where a 38 kHz subcarrier is used to separate the left-right difference signal from the central left-right sum channel, prior to the frequency modulation of the composite signal. An analog NTSC television channel is divided into subcarrier frequencies for video, color, and audio. DSL uses different frequencies for voice and for upstream and downstream data transmission on the same conductors, which is also an example of frequency duplex.

Where frequency-division multiplexing is used as to allow multiple users to share a physical communications channel, it is called frequency-division multiple access (FDMA).

FDMA is the traditional way of separating radio signals from different transmitters.

In the 1860s and 70s, several inventors attempted FDM under the names of acoustic telegraphy and harmonic telegraphy. Practical FDM was only achieved in the electronic age. Meanwhile, their efforts led to an elementary understanding of electroacoustic technology, resulting in the invention of the telephone.

CARRIER SYSTEMS

T1 Carrier System

T1 digital carrier system is a North American digital multiplexing standard since 1963. T1 stands for transmission one and specifies a digital carrier system using PCM encoded analog signal.

A T1 carrier system is time division multiplexes PCM encoded samples from 24 voice band channels for transmission over a single metallic wire pair or optical fiber transmission line. Each voice band channel has BW around 300Hz to 3000KHz.

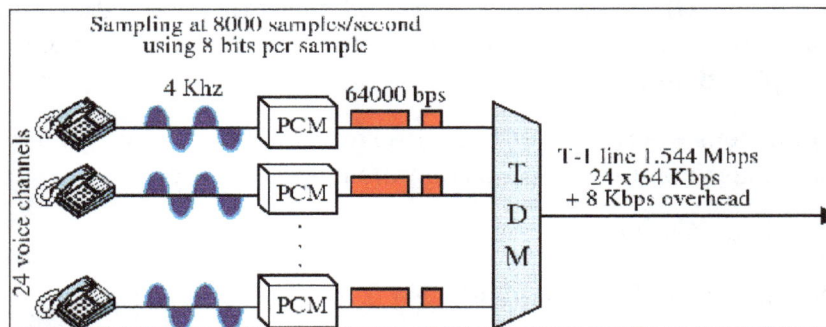

T1 Digital System.

A multiplexer is simply a digital switch with 24 independent inputs and one time division multiplexed output. The PCM output signals from 24 voice band channels are sequentially selected and connected through the multiplexer to the transmission line. With T1 carrier system, there is sampling, encoding and multiplexing of 24 voice band channels. Each channel contains an 8-bit PCM code and sampled 8000 times a second. Each channel is sampled at same rate but not at same time. The figure shows that, each channel is sampled once in each frame, but not at same time. Each channel's sample is offset from previous channel's sample by 1/24 of total frame time. Therefore one 64Kbps PCM encoded sample is transmitted for each voice band channel during each frame. The line Speed is calculated as:

$$\frac{24 \text{ Channels}}{\text{Frame}} \times \frac{8 \text{ Bits}}{\text{Channel}} = 192 \text{bits per frame}$$

Thus,

$$\frac{192\,\text{bits}}{\text{Frame}} \times \frac{8000\,\text{frames}}{\text{second}} = 1.536\,\text{Mbps}$$

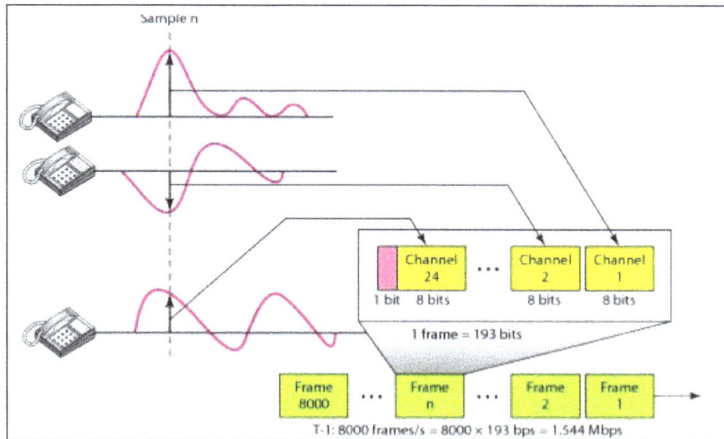

T-1 Frame Structure.

An additional bit (called framing bit) is added to each frame. The framing bit occurs once per frame (8000bps rate) and recovered in receiver, where it is used to maintain frame and sample synchronization between TDM transmitter and receiver. So each frame contains 193 bits and line speed for T1 digital carrier system is,

$$\frac{193\,\text{bits}}{\text{Frame}} \times \frac{8000\,\text{frames}}{\text{second}} = 1.544\,\text{Mbps}$$

AMI line coding is used for T1 digital Systems.

E Carrier System

The E-carrier is a member of the series of carrier systems developed for digital transmission of many simultaneous telephone calls by time-division multiplexing. The European Conference of Postal and Telecommunications Administrations (CEPT) originally standardized the E-carrier system, which revised and improved the earlier American T-carrier technology, and this has now been adopted by the International Telecommunication Union Telecommunication Standardization Sector (ITU-T). It was widely adopted in almost all countries outside the US, Canada, and Japan. E-carrier deployments have steadily been replaced by Ethernet as telecommunication networks transitions towards all IP.

E1 frame Structure

An E1 link operates over two separate sets of wires, usually unshielded twisted pair (balanced cable) or using coaxial (unbalanced cable). A nominal 3 volt peak signal is encoded with pulses using a method avoiding long periods without polarity changes. The line data rate is 2.048 Mbit/s (full duplex, i.e. 2.048 Mbit/s downstream and 2.048 Mbit/s upstream) which is split into 32 timeslots, each being allocated 8 bits in turn. Thus each timeslot sends and receives an 8-bit PCM sample, usually encoded according to A-law algorithm, 8,000 times per second ($8 \times 8,000 \times 32 = 2,048,000$). This is ideal for voice telephone calls where the voice is sampled at that data rate and reconstructed at the other end. The timeslots are numbered from 0 to 31.

The E1 frame defines a cyclical set of 32 time slots of 8 bits. The time slot 0 is devoted to transmission management and time slot 16 for signaling; the rest were assigned originally for voice/data transport.

The main characteristics of the 2-Mbit/s frame are described in the following figure.

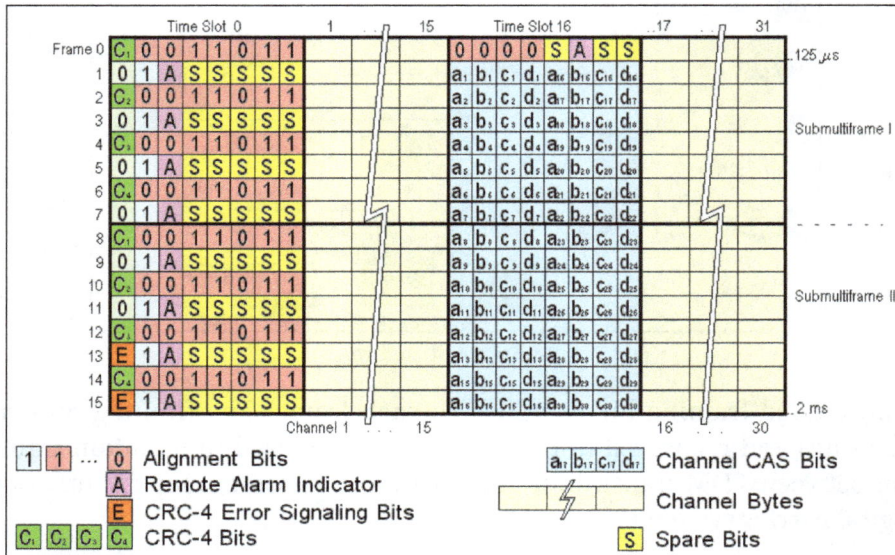

Special Timeslots

One timeslot (TS0) is reserved for framing purposes, and alternately transmits a fixed pattern. This allows the receiver to lock onto the start of each frame and match up each channel in turn. The standards allow for a full cyclic redundancy check to be performed across all bits transmitted in each frame, to detect if the circuit is losing bits (information), but this is not always used. An alarm signal may also be transmitted using timeslot TS0. Finally, some bits are reserved for national use.

One timeslot (TS16) is often reserved for signalling purposes, to control call setup and teardown according to one of several standard telecommunications protocols. This includes channel-associated signaling (CAS) where a set of bits is used to replicate opening and closing the circuit (as if picking up the telephone receiver and pulsing digits on a rotary phone), or using tone signalling which is passed through on the voice circuits themselves. More recent systems use common-channel signaling (CCS) such Signalling System 7 (SS7) where no particular timeslot is reserved for signalling purposes, the signalling protocol being transmitted on a freely chosen set of timeslots or on a different physical channel.

Frame Alignment

In an E1 channel, communication consists of sending consecutive frames from the transmitter to the receiver. The receiver must receive an indication showing when the first interval of each frame begins, so that, since it knows to which channel the information in each time slot corresponds, it can demultiplex correctly. This way, the bytes received in each slot are assigned to the correct channel. A synchronization process is then established, and it is known as frame alignment.

Frame-alignment Signal

In order to implement the frame alignment system so that the receiver of the frame can tell where it begins, there is so called a frame alignment signal (FAS). In the 2 Mbit/s frame system, the FAS is a combination of seven fixed bits ("0011011") transmitted in the first time slot in the frame (time slot zero or TS0). For the alignment mechanism to be maintained, the FAS does not need to be transmitted in every frame. Instead, this signal can be sent in alternate frames (in the first, in the third, in the fifth, and so on). In this case, TS0 is used as the synchronization slot. The TS0 of the rest of the frames is therefore available for other functions, such as the transmission of the alarms.

Multiframe CRC-4

In the TS0 of frames with FAS, the first bit is dedicated to carrying the cyclic redundancy checksum (CRC). It tells us whether there are one or more bit errors in a specific group of data received in the previous block of eight frames known as submultiframe.

The CRC-4 Procedure

The aim of this system is to avoid loss of synchronization due to the coincidental appearance of the sequence "0011011" in a time slot other than the TS0 of a frame with FAS. To implement the CRC code in the transmission of 2 Mbit/s frames, a CRC-4 multiframe is built, made up of 16 frames. These are then grouped in two blocks of eight frames called submultiframes, over which a CRC checksum or word of four bits (CRC-4) is put in the positions Ci (bits #1, frames with FAS) of the next submultiframe.

At the receiving end, the CRC of each submultiframe is calculated locally and compared to the CRC value received in the next submultiframe. If these do not coincide, one or more bit errors is determined to have been found in the block, and an alarm is sent back to the transmitter, indicating that the block received at the far end contains errors.

CRC-4 Multiframe Alignment

The receiving end has to know which is the first bit of the CRC-4 word (C_1). For this reason, a CRC-4 multiframe alignment word is needed. Obviously, the receiver has to be told where the multiframe begins (synchronization). The CRC-4 multiframe alignment word is the set combination "001011", which is introduced in the first bits of the frames that do not contain the FAS signal.

Advantages of the CRC-4 Method

The CRC-4 method is mainly used to protect the communication against a wrong frame alignment word, and also to provide a certain degree of monitoring of the bit error rate (BER), when this has low values (around 10^{-6}). This method is not suitable for cases in which the BER is around 10^{-3} (where each block contains at least one errored bit).

Another advantage in using the CRC is that all the bits transmitted are checked, unlike those systems that only check seven bits (those of the FAS, which are the only ones known in advance) out of every 51 bits (those between one FAS and the next). However, the CRC-4 code is not completely infallible, since there is a probability of around $\frac{1}{16}$ that an error may occur and not be detected; that is, that 6.25% of the blocks may contain errors that are not detected by the code.

Monitoring Errors

The aim of monitoring errors is to continuously check transmission quality without disturbing the information traffic and, when this quality is not of the required standard, taking the necessary steps to improve it. Telephone traffic is two way, which means that information is transmitted in both directions between the ends of the communication. This in its turn means that two 2 Mbit/s channels and two directions for transmission must be considered.

The CRC-4 multiframe alignment word only takes up six of the first eight bits of the TS0 without FAS. There are two bits in every second block or submultiframe, whose task is to indicate block errors in the far end of the communication. The mechanism is as follows: Both bits (called E-bits) have "1" as their default value. When the far end of the communication receives a 2 Mbit/s frame and detects an erroneous block, it puts a "0" in the E-bit that corresponds to the block in the frame being sent along the return path to the transmitter. This way, the near end of the communication is informed that an erroneous block has been detected, and both ends have the same information: one from the CRC-4 procedure and the other from the E bits. If we number the frames in the multiframe from 0 to 15, the E-bit of frame 13 refers to the submultiframe I (block I) received at the far end, and the E-bit of frame 15 refers to the submultiframe II (block II).

Supervision Bits

The bits that are in position 2 of the TS0 in the frame that does not contain the FAS are called supervision bits and are set to "1," to avoid simulations of the FAS signal.

NFASs: Spare Bits

The bits of the TS0 that do not contain the FAS in positions 3–8 make up what is known as the non-frame alignment signal or NFAS. This signal is sent in alternate frames (frame 1, frame 3, frame 5, etc.). The first bit of the NFAS (bit 3 of the TS0) is used to indicate that an alarm has occurred at the far end of the communication. When operating normally, it is set to "0", while a value of "1" indicates an alarm.

The bits in positions 4–8 are spare bits), and they do not have one single application, but can be used in a number of ways, as decided by the telecommunications carrier. In accordance with the ITU-T Rec. G.704, these bits can be used in specific point-to-point applications, or to establish a data link based on messages for operations management, maintenance or monitoring of the transmission quality, and so on. If these spare bits in the NFAS are not used, they must be set to "1" in international links.

NFAS: Alarm Bit

The method used to transmit the alarm makes use of the fact that in telephone systems, transmission is always two way). Multiplexing/demultiplexing devices (known generically as multiplex devices) are installed at both ends of the communication for the transmission and reception of frames. An alarm must be sent to the transmitter when a device detects either a power failure or a failure of the coder/decoder, in its multiplexer; or any of the following in its demultiplexer: loss of the signal (LOS), loss of frame alignment (LOF), or a BER greater than 10^{-3}.

The remote alarm indication (RAI) is sent in the NFAS of the return frames, with bit 3 set to "1". The transmitter then considers how serious the alarm is, and goes on generating a series of operations, depending on the type of alarm condition detected.

Signaling Channel

As well as transmitting information generated by the users of a telephone network, it is also necessary to transmit signaling information. Signaling refers to the protocols that must be established between exchanges so that the users can exchange information between them.

There are signals that indicate when a subscriber has picked up the telephone, when he or she can start to dial a number, and when another subscriber calls, as well as signals that let the communication link be maintained, and so on. In the E1 PCM system, signaling information can be transmitted by two different methods: the common channel signaling (CCS) method and the channel associated signaling (CAS) method. In both cases, the time slot TS16 of the basic 2 Mbit/s frame is used to transmit the signaling information.

For CCS signaling, messages of several bytes are transmitted through the 64 kbit/s channel provided by the TS16 of the frame, with these messages providing the signaling for all the channels in the frame. Each message contains information that determines the channel that is signaling. The signaling circuits access the 64 kbit/s channel of the TS16, and they are also common to all the channels signaled. There are different CCS systems that constitute complex protocols. In the following section and by way of example, channel associated signaling will be looked.

The Plesiochronous Digital Hierarchy

Based on the E1 signal, the ITU defined a hierarchy of plesiochronous signals that enables signals to be transported at rates of up to 140 Mbit/s. This section describes the characteristics of this hierarchy and the mechanism for dealing with fluctuations in respect to the nominal values of these rates, which are produced as a consequence of the tolerances of the system.

Higher Hierarchical Levels

As is the case with level 1 of the plesiochronous digital hierarchy (2 Mbit/s), the higher levels of multiplexing are carried out bit by bit (unlike the multiplexing of 64 kbit/s channels in a 2 Mbit/s signal, which is byte by byte), thus making it impossible to identify the lower level frames inside a higher level frame. Recovering the tributary frames requires the signal to be fully demultiplexed.

The higher hierarchical levels (8,448, 34,368, and 139,264 Kbit/s, etc.; referred to as 8, 34, and 140 Mbit/s for simplicity) are obtained by multiplexing four lower level frames within a frame whose nominal transmission rate is more than four times that of the lower level, in order to leave room for the permitted variations in rate (justification bits), as well as the corresponding FAS, alarm, and spare bits.

Multiplexing Level 2: 8 Mbit/s

The 8 Mbit/s frame structure is defined in the ITU-T Rec. G.742. The frame is divided into four groups, each of length 212 bits:

- Group I contains the FAS, with sequence "1111010000"; the A-bit (remote alarm); the S-bit (spare); and 200 T-bits (tributary) transporting data.

- Groups II and III contain a block of four J-bits (justification control) and 208 T-bits transporting data.

- Group IV contains a block of four J-bits, a block of R-bits (justification opportunity), one per tributary, and 204 T-bits. To check whether R-bits have been used, the J-bits are analyzed in each of the groups II, III, and IV (there are three per tributary). Ideally the R-bit does not carry useful information on 42.4% of the occasions. In other words, this percentage is the probability of justification or the insertion of stuffing bits.

212 Bits/Row				
Group	Bits			
I	FAS 1111010000 (10 bits)	A	S	200 Tributary Bits
II	J1 | J2 | J3 | J4	208 Tributary Bits		
III	J1 | J2 | J3 | J4	208 Tributary Bits		
IV	J1 | J2 | J3 | J4	R1 | R2 | R3 | R4	204 Tributary Bits	

Multiplexing Level 3: 34 Mbit/s

As in the previous case, the frame is divided into four groups, each of length 384 bits:

- Group I contains the FAS, with sequence "1111010000"; the A-bit (remote alarm); the S-bit (spare); and 372 T-bits (tributary) transporting data.

- Groups II and III contain a block of four J-bits (justification control) and 380 T-bits transporting data.

- Group IV contains a block of four J-bits, a block of R-bits (justification opportunity) one per tributary, and 376 T-bits. To check whether R-bits have been used, the J-bits are analyzed in each of the groups II, III, and IV (there are three per tributary). Ideally the R-bit does not carry useful information on 43.6% of the occasions.

384 Bits/Row				
Group	Bits			
I	FAS 1111010000 (10 bits)	A	S	372 Tributary Bits
II	J1 | J2 | J3 | J4	380 Tributary Bits		
III	J1 | J2 | J3 | J4	380 Tributary Bits		
IV	J1 | J2 | J3 | J4	R1 | R2 | R3 | R4	376 Tributary Bits	

Multiplexing Level 4: 140 Mbit/s

In this case, the frame is divided into six groups, each of length 488 bits:

- Group I contains the FAS, with sequence "111110100000"; the A-bit (remote alarm); three S-bits (spare); and 472 T-bits (tributary) transporting data.

- Groups II, III, IV, and V contain a block of four J-bits (justification control) and 484 T-bits transporting data.

- Group VI contains a block of four J-bits, a block of R-bits (justification opportunity), one per tributary, and 480 T-bits. To check whether R-bits have been used, the J-bits are analyzed in each of the groups II, III, IV, V, and VI (there are five per tributary). Ideally the R-bit does not carry useful information on 41.9% of the occasions.

488 Bits/Row				
Group	Bits			
I	FAS 111110100000 (12 bits)	A	3 S Bits	472 Tributary Bits
II	J1 | J2 | J3 | J4	484 Tributary Bits		
III	J1 | J2 | J3 | J4	484 Tributary Bits		
IV	J1 | J2 | J3 | J4	484 Tributary Bits		
V	J1 | J2 | J3 | J4	484 Tributary Bits		
VI	J1 | J2 | J3 | J4	R1 | R2 | R3 | R4	480 Tributary Bits	

The PDH hierarchy, with four levels from 2 to 140 Mbit/s.							
Level	Standard	Rate	Size	Frame/s	Code	Amplitude	Attenuation
E1	G.704/732	2.048 Mbit/s ± 50 ppm	256 bits	8,000	HDB3	2.37-3.00 V	6 dB
E2	G.742	8.448 Mbit/s ± 30 ppm	848 bits	9,962.2	HDB3	2.37 V	6 dB
E3	G.751	34.368 Mbit/s ± 20 ppm	1536 bits	22,375.0	HDB3	1.00 V	12 dB
E4	G.751	139.264 Mbit/s ± 15 ppm	2928 bits	47,562.8	CMI	1.00 V	12 dB

Service Bits in Higher-level Frames

In any of the groups containing the FAS in the 8, 34, and 140 Mbit/s frames, alarm bits and spare bits are also to be found. These are known as service bits. The A-bits (alarm) carry an alarm indication to the remote multiplexing device, when certain breakdown conditions are detected in the near-end device. The spare bits are designed for national use, and must be set to "1" in digital paths that cross international boundaries.

Plesiochronous Synchronization

As far as synchronization is concerned, the multiplexing of plesiochronous signals is not completely trouble free, especially when it comes to demultiplexing the circuits. In a PCM multiplexer of 30 + 2 channels, a sample of the output signal clock (1/32) is sent to the coders, so that the input channels are synchronized with the output frame. However, higher level multiplexers receive frames from lower level multiplexers with clocks whose value fluctuates around a nominal frequency value within certain margins of tolerance.

The margins are set by the ITU-T recommendations for each hierarchical level. The signals thus formed are almost synchronous, except for differences within the permitted margins of tolerance, and for this reason they are called plesiochronous.

Positive Justification

In order to perform bit-by-bit TDM, each higher-order PDH multiplexer has elastic memories in each of its inputs in which the incoming bits from each lower level signal line or tributary are written. Since the tributary signals have different rates, they are asynchronous with respect to each other. To prevent the capacity of the elastic memories from overflowing, the multiplexer reads the incoming bits at the maximum rate permitted within the range of tolerances.

When the rate of the incoming flow in any of the tributary lines is below this reading rate, the multiplexer cannot read any bits from the elastic memory, and so it uses a stuffing bit or justification bit (called justification opportunity) in the output aggregate signal. Its task is that of adapting the signal that enters the multiplexer to the rate at which this signal is transmitted within the output frame (its highest clock value). This type of justification is called positive justification. Justification bits, together with other overhead bits, make the output rate higher than the total of the input signals.

Justification Opportunity Bits

The task of the justification opportunity bits (R-bits) is to be available as extra bits that can be used when the rate of the incoming tributaries is higher than its nominal value (within the margin specified by ITU-T) by an amount that makes this necessary. In this case, the opportunity bit is no longer mere stuffing, but becomes an information bit instead.

In order for the device that receives the multiplexed signal to be able to determine whether a justification opportunity bit contains useful information (i.e. information from a tributary), justification control bits (J-bits) are included in the frame. Each group of control bits refers to one of the tributaries of the frame. All of them will be set to "0" if the associated opportunity bit is carrying useful information; otherwise they will be set to "1".

Several bits are used instead of just one, to provide protection against possible errors in transmission. On examining the control bits received, if they do not all have the same value, it is decided that they were sent with the majority value (a "1" if there are more 1s than 0s, for instance; it is assumed that there has been an error in the bits that are at 0).

It can be seen that there is a dispersion of the control bits referring to a tributary that causes them to be in separate groups. Spreading out the J-bits (control bits), reduces the probability of errors occurring in them, and a wrong decision being made as to whether or not they have been used as a useful data bit. If the wrong decision is made, there is not only an error in the output data, but also a slip of one bit; that is, the loss or repetition of one bit of information.

Managing Alarms in Higher-level Hierarchies

The A-bit of the FAS in 8, 34, and 140 Mbit/s frames enables the multiplexers that correspond to these hierarchies to transmit alarm indications to the far ends when a multiplexer detects an alarm condition.

In addition, 140 Mbit/s multiplexers also transmit an alarm indication when faced with the loss of frame alignment of the 34-Mbit/s signals received inside the 140 Mbit/s signals, as well as in the NFAS of the 34 Mbit/s signal that has lost its alignment (bit 11 of group I changes from "0" to "1") in the return channel.

- Link – A unidirectional channel residing in one timeslot of an E1 or T1 Line, carrying 64 kbit/s (64,000 bit/s) raw digital data.

- Line – A unidirectional E1 or T1 physical connection.

- Trunk – A bidirectional E1 or T1 physical connection.

Hierarchy levels

The PDH based on the E0 signal rate is designed so that each higher level can multiplex a set of lower level signals. Framed E1 is designed to carry 30 or 31 E0 data channels plus 1 or 2 special channels, all other levels are designed to carry 4 signals from the level below. Because of the necessity for overhead bits, and justification bits to account for rate differences between sections of the network, each subsequent level has a capacity greater than would be expected from simply multiplying the lower level signal rate (so for example E2 is 8.448 Mbit/s and not 8.192 Mbit/s as one might expect when multiplying the E1 rate by 4).

References

- Pulse-amplitude-modulation: elprocus.com, Retrieved 14 July, 2019

- Yuichiro Fujiwara (2013). "Self-synchronizing pulse position modulation with error tolerance". IEEE Transactions on Information Theory. 59: 5352–5362. Arxiv:1301.3369. Doi:10.1109/TIT.2013.2262094

- Pulse-code-modulation-and-demodulation: elprocus.com, Retrieved 19 April, 2019

- White, Curt (2007). Data Communications and Computer Networks. Boston, MA: Thomson Course Technology. Pp. 143–152. ISBN 1-4188-3610-9

- Pulse-width-modulation, guide: openlabpro.com, Retrieved 17 May, 2019

- Guowang Miao; Jens Zander; Ki Won Sung; Ben Slimane (2016). Fundamentals of Mobile Data Networks. Cambridge University Press. ISBN 1107143217

- Differential-pulse-code-modulation-working-application: elprocus.com, Retrieved 25 Februray, 2019

3

Digital Carrier Modulation Techniques

The modulation techniques which use discrete signals to modulate a carrier wave are known as digital carrier modulation. This chapter has been carefully written to provide an easy understanding of the various types of digital carrier modulation techniques such as amplitude shift keying, frequency shift keying, phase shift keying, etc.

DIGITAL COMMUNICATION SYSTEM

Digital communication systems are communication systems that use digital sequence (typically binary digits, that is, bits) as an interface between the source coding part and the channel coding part.

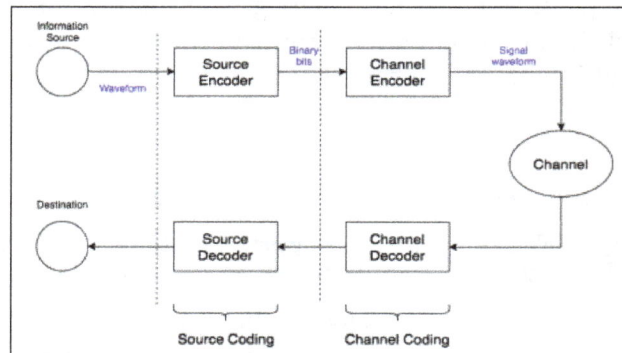

Basic Diagram of Digital Communication System.

- Source coding: The source encoder converts information waveforms to bits, while the decoder converts bits back to waveforms.

- Channel coding: The channel encoder converts bits to signal waveform, while the decoder converts received waveform back to bits.

Benefits of separating source and channel with a binary interface:

- Digital Hardware - cheap, reliable, scalable.

- Source/Channel Coding can be independently designed.

- Source-Channel Separation attains optimal transmission efficiency (Shannon's Theorem):

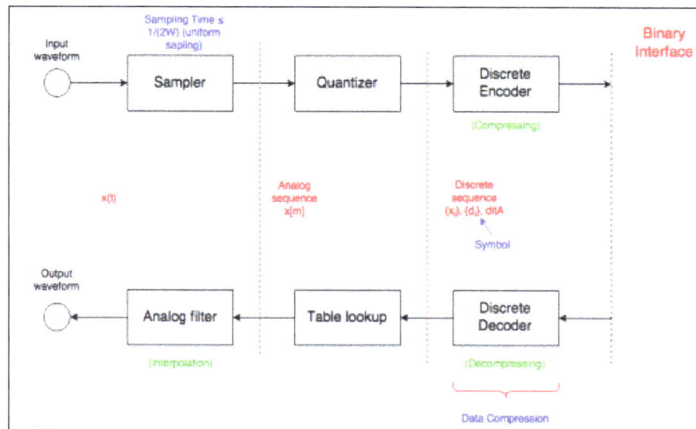

Basic Diagram of Source Coding.

Source Coding

Source encoding aims to convert information waveforms (text, audio, image, video, etc.) into bits, the universal currency of information in the digital world. The three major steps are:

- Sampling: Convert the continuous-time analog waveform to discrete-time sequence (but still continuous-valued).

- Quantization: Convert each continuous-valued symbol to discrete-valued representatives.

- Data compression: Remove the redundancy in the data and generate roughly i.i.d. uniformly distributed bits.

Source decoding does the reverse of encoding.

Channel Coding

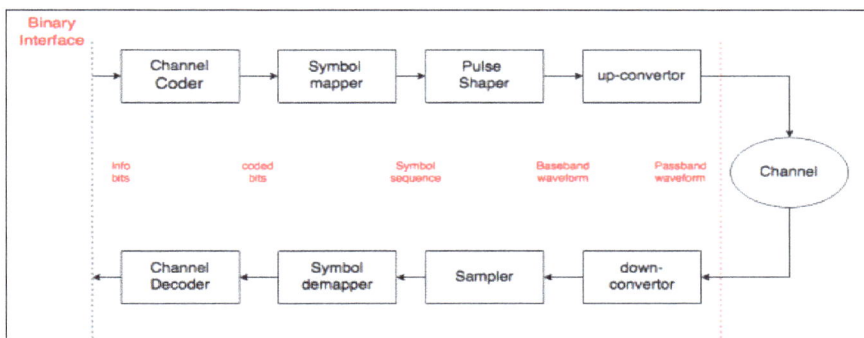

Basic Diagram of Channel Coding.

Channel encoding aims to convert information bits into passband waveformsbits, the universal currency of information in the digital world. The four major steps are:

- Error correcting codes: introduce redundancy into the information bits and produce longer coded bits. Remark Example of "error correction":

  – Repetition code: each bit repeat N times.

Channel noise: flip the bit $n.p.\,p$, w.$p.(1 - p)$ remain the same.

    – Bit Error happens when there are more than $\dfrac{N}{2}$ bit flips.

$$\text{Pr\{error\}}= 1-\sum_{i=0}^{\left[\frac{N}{2}\right]}\binom{N}{i}p^{i}(1-p)^{n-i}$$

- Symbol mapping: Map the coded bits to constellation points, each of which is a complex symbol.

- Pulse shaping: Modulate the symbol to suitable baseband waveforms. There are some specific conditions needed to be satisfied, which will be discussed in later lectures.

- Up conversion: Convert the baseband waveform to passband waveform, so that the effective frequency band follows the constraints from the physical world.

Channel decoding does the reverse of encoding.

From Analog to Digital and Back

Two key elements in digital communication systems can be recognized:

- Conversion between (discrete-time) sequences and (continuous-time) waveforms,

$$\begin{cases} \text{source coding: sampling/interpolation filter} \\ \text{channel coding: pulse shaping/sampling} \end{cases}$$

- Conversion between bits and symbol sequences,

$$\begin{cases} \text{source coding: quantization/table lookup} \\ \text{channel coding: symbol mapper/demapper} \end{cases}$$

You can find the similarity and differences between source and channel coding.

Conversion between Sequences and Waveforms via Orthogonal Expansion

We have learned two approaches in signals and systems,

- For time-limited signals $x(t) : 0 \le t \le T$ Fourier Series.

- For band-limited signals $x(f) : -W \le f \le W$ Sampling Theorem (Nyquist–Shannon sampling theorem).

	Analysis $\left(x(t) \to x[m]\right)$	*Sythesis* $\left(x[m] \to x(t)\right)$
Fourier Series	$x(m) = \dfrac{I}{T}\int_{T} x(t)e^{-\frac{j2\pi m}{T}t}\,dt$	$x(t) = \displaystyle\sum_{m=-\infty}^{\infty} x[m]\,e^{\frac{j\,2\pi m}{T}t}$
Sampling Theorem	$x[m] = x\left(m \cdot \dfrac{1}{2W}\right)$	$x(t) = \displaystyle\sum_{m=-\infty}^{\infty} x[m]\text{-sinc}(2w_{t}\text{-}m)$

$$\text{sinc}(t) \triangleq \frac{\sin(\pi t)}{\pi t}$$

Signal Space Point of View

- Synthesis:

$$\{x[m]\} \rightarrow \{\phi m\} \;\rightarrow\; x(t) \;=\; \sum_{m=-\infty}^{\infty} x[m] \cdot \phi m(t)$$

From a linear algebra point of view,

$$\text{vector} \longleftrightarrow \text{waveform}$$

$$\text{basis} \longleftrightarrow \{\phi_m(t)\}$$

The basis of the functional space:

Time-limited signals: $\Phi \;=\; \{\phi m(t)\} \;=\; \left\{ e^{j\frac{2\pi m}{T}t},\; m \in Z \right\}$

band-limited signals: $\Phi \;=\; \{ sinc(2W\,t - m),\; m \in Z \}$

In practice: one usually choose $\phi_m(t) \;=\; p(t - mT)$, where,

p($\cdot$) : "Pulse" in Pulse Shaping,

$$x(t) \;=\; \sum_{m=-\infty}^{+\infty} x[m]p(t - mT).$$

Example:

1. Sinc pulse: $p(t) \;=\; sinc(2Wt),\; W = \dfrac{1}{T}$

2. Raised cosine: $p(t) = sinc\dfrac{t}{T} \cdot \dfrac{cos\left(\dfrac{\pi\beta t}{T}\right)}{1 - \dfrac{4\beta t^2}{T^2}}$

- Analysis (Decomposition):

$$x(t) \;\rightarrow\; \phi_m(t) \rightarrow x[m]$$

If $\{\phi_m\}$ forms an orthnormal basis, then $x[m] \;=<\; x(t),\; \phi_m(t)\; >$

- Orthonormal:

$$< \phi_i,\; \phi_j > = \begin{Bmatrix} 0 \; if \; i \; \neq j \\ 1 \; if \; i \; = j \end{Bmatrix}$$

– In the signal space, inner product is defined by "integral".

$$< \phi_i, \phi_j > \triangleq \int_{-\infty}^{\infty} \phi_i(\tau)\phi*_j(\tau)d\tau$$

AMPLITUDE SHIFT KEYING

Amplitude-shift keying (ASK) is a form of amplitude modulation that represents digital data as variations in the amplitude of a carrier wave. In an ASK system, the binary symbol 1 is represented by transmitting a fixed-amplitude carrier wave and fixed frequency for a bit duration of T seconds. If the signal value is 1 then the carrier signal will be transmitted; otherwise, a signal value of 0 will be transmitted.

Any digital modulation scheme uses a finite number of distinct signals to represent digital data. ASK uses a finite number of amplitudes, each assigned a unique pattern of binary digits. Usually, each amplitude encodes an equal number of bits. Each pattern of bits forms the symbol that is represented by the particular amplitude. The demodulator, which is designed specifically for the symbol-set used by the modulator, determines the amplitude of the received signal and maps it back to the symbol it represents, thus recovering the original data. Frequency and phase of the carrier are kept constant.

Like AM, an ASK is also linear and sensitive to atmospheric noise, distortions, propagation conditions on different routes in PSTN, etc. Both ASK modulation and demodulation processes are relatively inexpensive. The ASK technique is also commonly used to transmit digital data over optical fiber. For LED transmitters, binary 1 is represented by a short pulse of light and binary 0 by the absence of light. Laser transmitters normally have a fixed "bias" current that causes the device to emit a low light level. This low level represents binary 0, while a higher-amplitude lightwave represents binary 1.

The simplest and most common form of ASK operates as a switch, using the presence of a carrier wave to indicate a binary one and its absence to indicate a binary zero. This type of modulation is called on-off keying (OOK), and is used at radio frequencies to transmit Morse code (referred to as continuous wave operation),

More sophisticated encoding schemes have been developed which represent data in groups using additional amplitude levels. For instance, a four-level encoding scheme can represent two bits with each shift in amplitude; an eight-level scheme can represent three bits; and so on. These forms of amplitude-shift keying require a high signal-to-noise ratio for their recovery, as by their nature much of the signal is transmitted at reduced power.

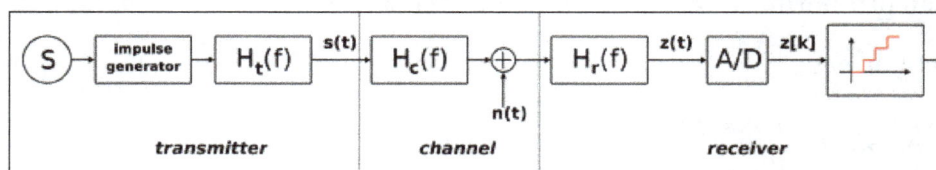

ASK diagram.

ASK system can be divided into three blocks. The first one represents the transmitter, the second one is a linear model of the effects of the channel, the third one shows the structure of the receiver. The following notation is used:

- $h_t(f)$ is the carrier signal for the transmission.

- $h_c(f)$ is the impulse response of the channel.

- $n(t)$ is the noise introduced by the channel.

- $h_r(f)$ is the filter at the receiver.

- L is the number of levels that are used for transmission.

- T_s is the time between the generation of two symbols.

Different symbols are represented with different voltages. If the maximum allowed value for the voltage is A, then all the possible values are in the range [−A, A] and they are given by:

$$v_i = \frac{2A}{L-1} i - A; \quad i = 0, 1, \ldots, L-1$$

the difference between one voltage and the other is:

$$\Delta = \frac{2A}{L-1}$$

Considering the picture, the symbols v[n] are generated randomly by the source S, then the impulse generator creates impulses with an area of v[n]. These impulses are sent to the filter ht to be sent through the channel. In other words, for each symbol a different carrier wave is sent with the relative amplitude.

Out of the transmitter, the signal s(t) can be expressed in the form:

$$s(t) = \sum_{n=-\infty}^{\infty} v[n] \cdot h_t(t - nT_s)$$

In the receiver, after the filtering through hr (t) the signal is:

$$z(t) = n_r(t) + \sum_{n=-\infty}^{\infty} v[n] \cdot g(t - nT_s)$$

where we use the notation:

$$n_r(t) = n(t) * h_r(t)$$
$$g(t) = h_t(t) * h_c(t) * h_r(t)$$

where * indicates the convolution between two signals. After the A/D conversion the signal z[k] can be expressed in the form:

$$z[k] = n_r[k] + v[k]g[0] + \sum_{n \neq k} v[n]g[k-n]$$

In this relationship, the second term represents the symbol to be extracted. The others are unwanted: the first one is the effect of noise, the third one is due to the intersymbol interference.

If the filters are chosen so that g(t) will satisfy the Nyquist ISI criterion, then there will be no intersymbol interference and the value of the sum will be zero, so:

$$z[k] = n_r[k] + v[k]g[0]$$

the transmission will be affected only by noise.

Probability of Error

The probability density function of having an error of a given size can be modelled by a Gaussian function; the mean value will be the relative sent value, and its variance will be given by:

$$\sigma_N^2 = \int_{-\infty}^{+\infty} \Phi_N(f) \cdot |H_r(f)|^2 \, df$$

where $\Phi_N(f)$ is the spectral density of the noise within the band and Hr (f) is the continuous Fourier transform of the impulse response of the filter hr (f).

The probability of making an error is given by:

$$P_e = P_{e|H_0} \cdot P_{H_0} + P_{e|H_1} \cdot P_{H_1} + \cdots + P_{e|H_{L-1}} \cdot P_{H_{L-1}} = \sum_{k=0}^{L-1} P_{e|H_k} \cdot P_{H_k}$$

where, for example, P_{H_0} is the conditional probability of making an error given that a symbol v0 has been sent and P_{H_0} is the probability of sending a symbol v0.

If the probability of sending any symbol is the same, then:

$$P_{H_i} = \frac{1}{L}$$

If we represent all the probability density functions on the same plot against the possible value of the voltage to be transmitted, we get a picture like this (the particular case of is shown):

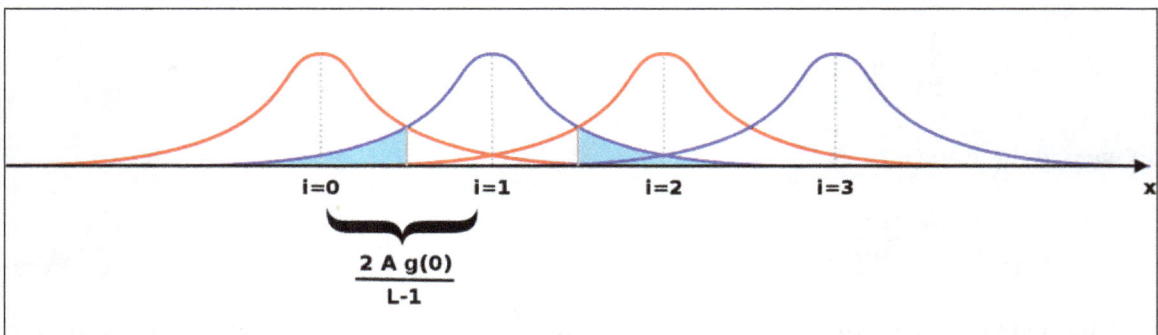

The probability of making an error after a single symbol has been sent is the area of the Gaussian function falling under the functions for the other symbols. It is shown in cyan for just one of them.

If we call P^+ the area under one side of the Gaussian, the sum of all the areas will be: $2LP^+ - 2P^+$. The total probability of making an error can be expressed in the form:

$$P_e = 2\left(1 - \frac{1}{L}\right)P^+$$

We now have to calculate the value of P^+. In order to do that, we can move the origin of the reference wherever we want: the area below the function will not change. We are in a situation like the one shown in the following image:

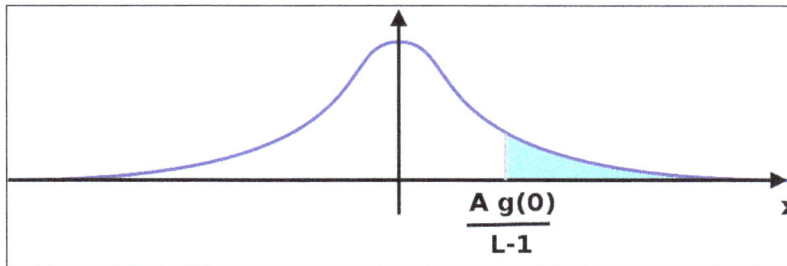

it does not matter which Gaussian function we are considering, the area we want to calculate will be the same. The value we are looking for will be given by the following integral:

$$P^+ = \int_{\frac{Ag(0)}{L-1}}^{\infty} \frac{1}{\sqrt{2\pi}\sigma_N} e^{-\frac{x^2}{2\sigma_N^2}} dx = \frac{1}{2}\mathrm{erfc}\left(\frac{Ag(0)}{\sqrt{2}(L-1)\sigma_N}\right)$$

where is the complementary error function. Putting all these results together, the probability to make an error is:

$$P_e = \left(1 - \frac{1}{L}\right)\mathrm{erfc}\left(\frac{Ag(0)}{\sqrt{2}(L-1)\sigma_N}\right)$$

from this formula we can easily understand that the probability to make an error decreases if the maximum amplitude of the transmitted signal or the amplification of the system becomes greater; on the other hand, it increases if the number of levels or the power of noise becomes greater.

This relationship is valid when there is no intersymbol interference, i.e. $g(t)$ is a Nyquist function.

FREQUENCY SHIFT KEYING

Frequency-shift keying (FSK) is a frequency modulation scheme in which digital information is transmitted through discrete frequency changes of a carrier signal. The technology is used for communication systems such as telemetry, weather balloon radiosondes, caller ID, garage door openers, and low frequency radio transmission in the VLF and ELF bands. The simplest FSK is binary FSK (BFSK). BFSK uses a pair of discrete frequencies to transmit binary (0s and 1s) information. With this scheme, the "1" is called the mark frequency and the "0" is called the space frequency.

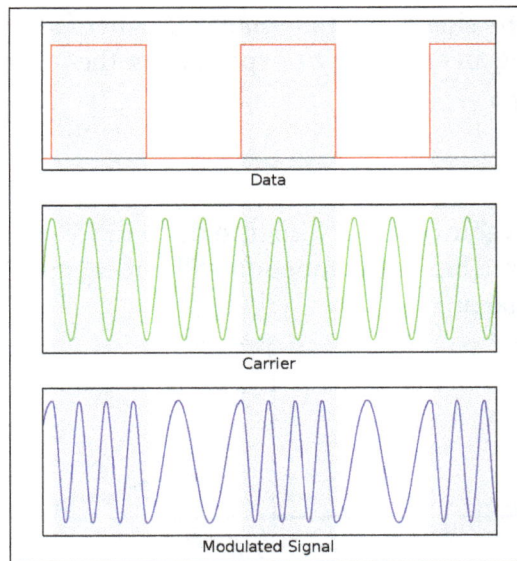

An example of binary FSK.

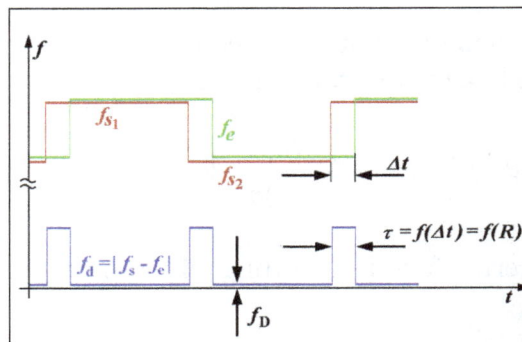

Modulating and Demodulating

Reference implementations of FSK modems exist and are documented in detail. The demodulation of a binary FSK signal can be done using the Goertzel algorithm very efficiently, even on low-power microcontrollers.

Other Forms of FSK

Continuous-phase Frequency-shift Keying

In principle FSK can be implemented by using completely independent free-running oscillators, and switching between them at the beginning of each symbol period. In general, independent oscillators will not be at the same phase and therefore the same amplitude at the switch-over instant, causing sudden discontinuities in the transmitted signal.

In practice, many FSK transmitters use only a single oscillator, and the process of switching to a different frequency at the beginning of each symbol period preserves the phase. The elimination of discontinuities in the phase (and therefore elimination of sudden changes in amplitude) reduces sideband power, reducing interference with neighboring channels.

Gaussian Frequency-shift Keying

Rather than directly modulating the frequency with the digital data symbols, "instantaneously" changing the frequency at the beginning of each symbol period, Gaussian frequency-shift keying (GFSK) filters the data pulses with a Gaussian filter to make the transitions smoother. This filter has the advantage of reducing sideband power, reducing interference with neighboring channels, at the cost of increasing intersymbol interference. It is used by DECT, Bluetooth, Cypress Wireless-sUSB, Nordic Semiconductor, Texas Instruments LPRF, Z-Wave and Wavenis devices. For basic data rate Bluetooth the minimum deviation is 115 kHz.

A GFSK modulator differs from a simple frequency-shift keying modulator in that before the baseband waveform (levels −1 and +1) goes into the FSK modulator, it is passed through a Gaussian filter to make the transitions smoother so to limit its spectral width. Gaussian filtering is a standard way for reducing spectral width; it is called "pulse shaping" in this application.

In ordinary non-filtered FSK, at a jump from −1 to +1 or +1 to −1, the modulated waveform changes rapidly, which introduces large out-of-band spectrum. If the pulse is changed going from −1 to +1 as −1, −0.98, −0.93, ..., +0.93, +0.98, +1, and this smoother pulse is used to determine the carrier frequency, the out-of-band spectrum will be reduced.

Minimum-shift Keying

Minimum frequency-shift keying or minimum-shift keying (MSK) is a particular spectrally efficient form of coherent FSK. In MSK, the difference between the higher and lower frequency is identical to half the bit rate. Consequently, the waveforms that represent a 0 and a 1 bit differ by exactly half a carrier period. The maximum frequency deviation is $\delta = 0.25 f_m$, where f_m is the maximum modulating frequency. As a result, the modulation index m is 0.5. This is the smallest FSK modulation index that can be chosen such that the waveforms for 0 and 1 are orthogonal.

Gaussian minimum-shift Keying

A variant of MSK called Gaussian minimum-shift keying (GMSK) is used in the GSM mobile phone standard.

Audio FSK

Audio frequency-shift keying (AFSK) is a modulation technique by which digital data is represented by changes in the frequency (pitch) of an audio tone, yielding an encoded signal suitable for transmission via radio or telephone. Normally, the transmitted audio alternates between two tones: one, the "mark", represents a binary one; the other, the "space", represents a binary zero.

AFSK differs from regular frequency-shift keying in performing the modulation at baseband frequencies. In radio applications, the AFSK-modulated signal normally is being used to modulate an RF carrier (using a conventional technique, such as AM or FM) for transmission.

AFSK is not always used for high-speed data communications, since it is far less efficient in both power and bandwidth than most other modulation modes. In addition to its simplicity, however,

AFSK has the advantage that encoded signals will pass through AC-coupled links, including most equipment originally designed to carry music or speech.

AFSK is used in the U.S.-based Emergency Alert System to notify stations of the type of emergency, locations affected, and the time of issue without actually hearing the text of the alert.

Continuous 4-level FM

Phase 1 radios in the Project 25 system use continuous 4-level FM (C4FM) modulation.

Applications

In 1910, Reginald Fessenden invented a two-tone method of transmitting Morse code. Dots and dashes were replaced with different tones of equal length. The intent was to minimize transmission time.

Some early CW transmitters employed an arc converter that could not be conveniently keyed. Instead of turning the arc on and off, the key slightly changed the transmitter frequency in a technique known as the *compensation-wave method*. The compensation-wave was not used at the receiver. Spark transmitters used for this method consumed a lot of bandwidth and caused interference, so it was discouraged by 1921.

Most early telephone-line modems used audio frequency-shift keying (AFSK) to send and receive data at rates up to about 1200 bits per second. The Bell 103 and Bell 202 modems used this technique. Even today, North American caller ID uses 1200 baud AFSK in the form of the Bell 202 standard. Some early microcomputers used a specific form of AFSK modulation, the Kansas City standard, to store data on audio cassettes. AFSK is still widely used in amateur radio, as it allows data transmission through unmodified voiceband equipment.

AFSK is also used in the United States' Emergency Alert System to transmit warning information. It is used at higher bitrates for Weathercopy used on Weatheradio by NOAA in the U.S.

The CHU shortwave radio station in Ottawa, Ontario, Canada broadcasts an exclusive digital time signal encoded using AFSK modulation.

Standards for use in Caller ID and remote metering

Frequency-shift keying (FSK) is commonly used over telephone lines for Caller ID (displaying callers' numbers) and remote metering applications. There are several variations of this technology.

European Telecommunications Standards Institute FSK

In some countries of Europe, the European Telecommunications Standards Institute (ETSI) standards 200 778-1 and -2 – replacing 300 778-1 & -2 – allow 3 physical transport layers (Telcordia Technologies (formerly Bellcore), British Telecom (BT) and Cable Communications Association (CCA)), combined with 2 data formats Multiple Data Message Format (MDMF) & Single Data Message Format (SDMF), plus the Dual-tone multi-frequency (DTMF) system and a no-ring mode for meter-reading and the like. It's more of a recognition that the different types exist than an attempt to define a single "standard".

Telcordia Technologies FSK

The Telcordia Technologies (formerly Bellcore) standard is used in the United States, Canada, Australia, China, Hong Kong and Singapore. It sends the data after the first ring tone and uses the 1200 bits per second Bell 202 tone modulation. The data may be sent in SDMF – which includes the date, time and number – or in MDMF, which adds a NAME field.

British Telecom FSK

British Telecom (BT) in the United Kingdom developed their own standard, which wakes up the display with a line reversal, then sends the data as CCITT v.23 modem tones in a format similar to MDMF. It is used by BT, wireless networks like the late Ionica, and some cable companies. Details are to be found in BT Supplier Information Notes (SINs) 227 and 242; another useful document is Designing Caller Identification Delivery Using XR-2211 for BT from the EXAR website.

Cable Communications Association FSK

The Cable Communications Association (CCA) of the United Kingdom developed their own standard which sends the information after a short first ring, as either Bell 202 or V.23 tones. They developed a new standard rather than change some "street boxes" (multiplexors) which couldn't cope with the BT standard. The UK cable industry use a variety of switches: most are Nortel DMS-100; some are System X; System Y; and Nokia DX220. Note that some of these use the BT standard instead of the CCA one. The data format is similar to the BT one, but the transport layer is more like Telcordia Technologies, so North American or European equipment is more likely to detect it.

PHASE SHIFT KEYING

Phase-shift keying (PSK) is a digital modulation process which conveys data by changing (modulating) the phase of a constant frequency reference signal (the carrier wave). The modulation is accomplished by varying the sine and cosine inputs at a precise time. It is widely used for wireless LANs, RFID and Bluetooth communication.

Any digital modulation scheme uses a finite number of distinct signals to represent digital data. PSK uses a finite number of phases, each assigned a unique pattern of binary digits. Usually, each phase encodes an equal number of bits. Each pattern of bits forms the symbol that is represented by the particular phase. The demodulator, which is designed specifically for the symbol-set used by the modulator, determines the phase of the received signal and maps it back to the symbol it represents, thus recovering the original data. This requires the receiver to be able to compare the phase of the received signal to a reference signal – such a system is termed coherent (and referred to as CPSK).

CPSK requires a complicated demodulator, because it must extract the reference wave from the received signal and keep track of it, to compare each sample to. Alternatively, the phase shift of each symbol sent can be measured with respect to the phase of the previous symbol sent. Because the symbols are encoded in the difference in phase between successive samples, this is called differential phase-shift keying (DPSK). DPSK can be significantly simpler to implement than ordinary PSK, as it is a 'non-coherent' scheme, i.e. there is no need for the demodulator to keep track of a reference wave. A trade-off is that it has more demodulation errors.

There are three major classes of digital modulation techniques used for transmission of digitally represented data:

- Amplitude-shift keying (ASK)

- Frequency-shift keying (FSK)

- Phase-shift keying (PSK)

All convey data by changing some aspect of a base signal, the carrier wave (usually a sinusoid), in response to a data signal. In the case of PSK, the phase is changed to represent the data signal. There are two fundamental ways of utilizing the phase of a signal in this way:

- By viewing the phase itself as conveying the information, in which case the demodulator must have a reference signal to compare the received signal's phase against; or

- By viewing the *change* in the phase as conveying information – *differential* schemes, some of which do not need a reference carrier (to a certain extent).

A convenient method to represent PSK schemes is on a constellation diagram. This shows the points in the complex plane where, in this context, the real and imaginary axes are termed the in-phase and quadrature axes respectively due to their 90° separation. Such a representation on perpendicular axes lends itself to straightforward implementation. The amplitude of each point along the in-phase axis is used to modulate a cosine (or sine) wave and the amplitude along the quadrature axis to modulate a sine (or cosine) wave. By convention, in-phase modulates cosine and quadrature modulates sine.

In PSK, the constellation points chosen are usually positioned with uniform angular spacing around a circle. This gives maximum phase-separation between adjacent points and thus the best immunity to corruption. They are positioned on a circle so that they can all be transmitted with the same energy. In this way, the moduli of the complex numbers they represent will be the same and thus so will the amplitudes needed for the cosine and sine waves. Two common examples are "binary phase-shift keying" (BPSK) which uses two phases, and "quadrature phase-shift keying" (QPSK) which uses four phases, although any number of phases may be used. Since the data to be conveyed are usually binary, the PSK scheme is usually designed with the number of constellation points being a power of two.

For determining error-rates mathematically, some definitions will be needed:

- E_b, energy per bit

- $E_s = nE_b$, energy per symbol with n bits

- T_b, bit duration

- T_s, symbol duration

- $\dfrac{1}{2}N_0$, noise power spectral density (W/Hz)

- P_b, probability of *bit-error*

- P_s, probability of symbol-error

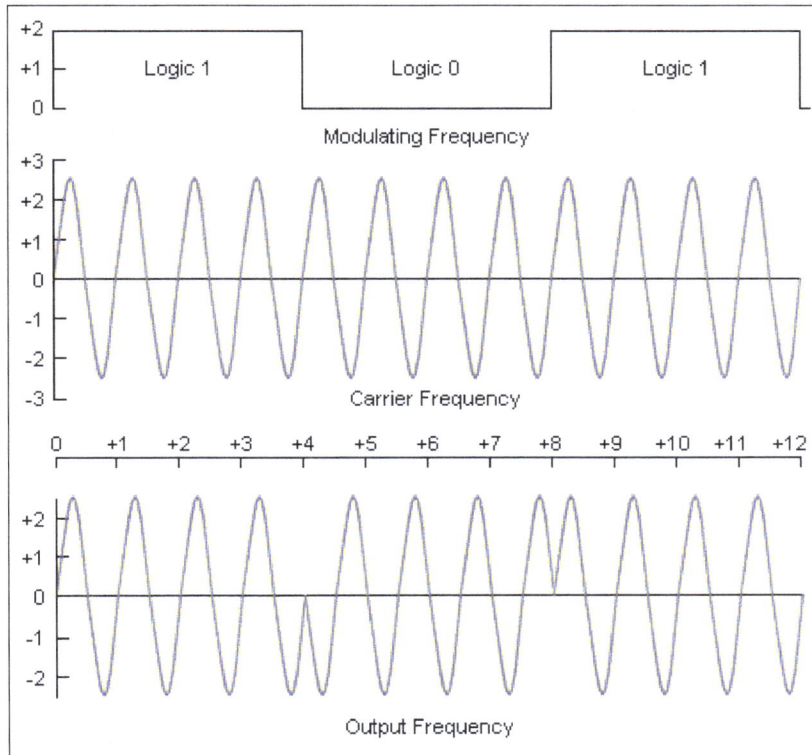

$Q(x)$ will give the probability that a single sample taken from a random process with zero-mean and unit-variance Gaussian probability density function will be greater or equal to x. It is a scaled form of the complementary Gaussian error function:

$$Q(x) = \frac{1}{\sqrt{2\pi}} \int_x^\infty e^{-\frac{1}{2}t^2} \, dt = \frac{1}{2}\mathrm{erfc}\left(\frac{x}{\sqrt{2}}\right), x \geq 0 \, .$$

The error rates quoted here are those in additive white Gaussian noise (AWGN). These error rates are lower than those computed in fading channels, hence, are a good theoretical benchmark to compare with.

Applications

Owing to PSK's simplicity, particularly when compared with its competitor quadrature amplitude modulation, it is widely used in existing technologies.

The wireless LAN standard, IEEE 802.11b-1999, uses a variety of different PSKs depending on the data rate required. At the basic rate of 1 Mbit/s, it uses DBPSK (differential BPSK). To provide the extended rate of 2 Mbit/s, DQPSK is used. In reaching 5.5 Mbit/s and the full rate of 11 Mbit/s,

QPSK is employed, but has to be coupled with complementary code keying. The higher-speed wireless LAN standard, IEEE 802.11g-2003, has eight data rates: 6, 9, 12, 18, 24, 36, 48 and 54 Mbit/s. The 6 and 9 Mbit/s modes use OFDM modulation where each sub-carrier is BPSK modulated. The 12 and 18 Mbit/s modes use OFDM with QPSK. The fastest four modes use OFDM with forms of quadrature amplitude modulation.

Because of its simplicity, BPSK is appropriate for low-cost passive transmitters, and is used in RFID standards such as ISO/IEC 14443 which has been adopted for biometric passports, credit cards such as American Express's ExpressPay, and many other applications.

Bluetooth 2 will use $\pi/4$-DQPSK at its lower rate (2 Mbit/s) and 8-DPSK at its higher rate (3 Mbit/s) when the link between the two devices is sufficiently robust. Bluetooth 1 modulates with Gaussian minimum-shift keying, a binary scheme, so either modulation choice in version 2 will yield a higher data-rate. A similar technology, IEEE 802.15.4 (the wireless standard used by ZigBee) also relies on PSK using two frequency bands: 868–915 MHz with BPSK and at 2.4 GHz with OQPSK.

Both QPSK and 8PSK are widely used in satellite broadcasting. QPSK is still widely used in the streaming of SD satellite channels and some HD channels. High definition programming is delivered almost exclusively in 8PSK due to the higher bitrates of HD video and the high cost of satellite bandwidth. The DVB-S2 standard requires support for both QPSK and 8PSK. The chipsets used in new satellite set top boxes, such as Broadcom's 7000 series support 8PSK and are backward compatible with the older standard.

Historically, voice-band synchronous modems such as the Bell 201, 208, and 209 and the CCITT V.26, V.27, V.29, V.32, and V.34 used PSK.

Binary Phase-shift Keying (BPSK)

BPSK (also sometimes called PRK, phase reversal keying, or 2PSK) is the simplest form of phase shift keying (PSK). It uses two phases which are separated by 180° and so can also be termed 2-PSK. It does not particularly matter exactly where the constellation points are positioned, and in this figure they are shown on the real axis, at 0° and 180°. Therefore, it handles the highest noise level or distortion before the demodulator reaches an incorrect decision. That makes it the most robust of all the PSKs. It is, however, only able to modulate at 1 bit/symbol and so is unsuitable for high data-rate applications.

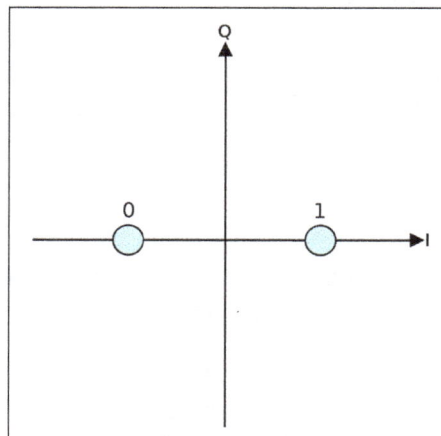

Constellation diagram example for BPSK.

In the presence of an arbitrary phase-shift introduced by the communications channel, the demodulator is unable to tell which constellation point is which. As a result, the data is often differentially encoded prior to modulation.

BPSK is functionally equivalent to 2-QAM modulation.

Implementation

The general form for BPSK follows the equation:

$$s_n(t) = \sqrt{\frac{2E_b}{T_b}} \cos(2\pi f t + \pi(1-n)), \quad n = 0,1.$$

This yields two phases, 0 and π. In the specific form, binary data is often conveyed with the following signals,

for binary 0,

$$s_0(t) = \sqrt{\frac{2E_b}{T_b}} \cos(2\pi f t + \pi) = -\sqrt{\frac{2E_b}{T_b}} \cos(2\pi f t)$$

for binary 1,

$$s_1(t) = \sqrt{\frac{2E_b}{T_b}} \cos(2\pi f t)$$

where f is the frequency of the base band.

Hence, the signal space can be represented by the single basis function,

$$\phi(t) = \sqrt{\frac{2}{T_b}} \cos(2\pi f t)$$

where 1 is represented by $\sqrt{E_b}\phi(t)$ and 0 is represented by $-\sqrt{E_b}\phi(t)$. This assignment is, of course, arbitrary.

This use of this basis function is shown at the end of the next section in a signal timing diagram. The topmost signal is a BPSK-modulated cosine wave that the BPSK modulator would produce. The bit-stream that causes this output is shown above the signal (the other parts of this figure are relevant only to QPSK). After modulation, the base band signal will be moved to the high frequency band by multiplying $\cos(2\pi f_c t)$.

Bit Error Rate

The bit error rate (BER) of BPSK under additive white Gaussian noise (AWGN) can be calculated as:

$$P_b = Q\left(\sqrt{\frac{2E_b}{N_0}}\right) \text{ or } P_e = \frac{1}{2}\text{erfc}\left(\sqrt{\frac{E_b}{N_0}}\right)$$

Since there is only one bit per symbol, this is also the symbol error rate.

Quadrature Phase-shift Keying (QPSK)

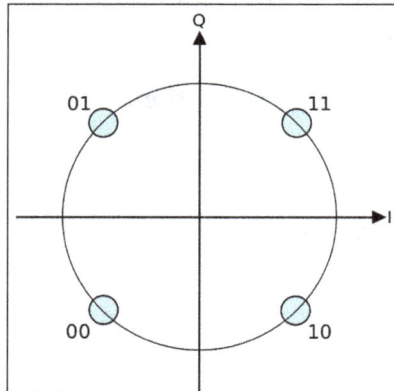

Constellation diagram for QPSK with Gray coding.
Each adjacent symbol only differs by one bit.

Sometimes this is known as *quadriphase PSK*, 4-PSK, or 4-QAM. (Although the root concepts of QPSK and 4-QAM are different, the resulting modulated radio waves are exactly the same.) QPSK uses four points on the constellation diagram, equispaced around a circle. With four phases, QPSK can encode two bits per symbol, shown in the diagram with Gray coding to minimize the bit error rate (BER) – sometimes misperceived as twice the BER of BPSK.

The mathematical analysis shows that QPSK can be used either to double the data rate compared with a BPSK system while maintaining the *same* bandwidth of the signal, or to *maintain the data-rate of BPSK* but halving the bandwidth needed. In this latter case, the BER of QPSK is *exactly the same* as the BER of BPSK – and deciding differently is a common confusion when considering or describing QPSK. The transmitted carrier can undergo numbers of phase changes.

Given that radio communication channels are allocated by agencies such as the Federal Communication Commission giving a prescribed (maximum) bandwidth, the advantage of QPSK over BPSK becomes evident: QPSK transmits twice the data rate in a given bandwidth compared to BPSK - at the same BER. The engineering penalty that is paid is that QPSK transmitters and receivers are more complicated than the ones for BPSK. However, with modern electronics technology, the penalty in cost is very moderate.

As with BPSK, there are phase ambiguity problems at the receiving end, and differentially encoded QPSK is often used in practice.

Implementation

The implementation of QPSK is more general than that of BPSK and also indicates the implementation of higher-order PSK. Writing the symbols in the constellation diagram in terms of the sine and cosine waves used to transmit them:

$$s_n(t) = \sqrt{\frac{2E_s}{T_s}} \cos\left(2\pi f_c t + (2n-1)\frac{\pi}{4}\right), \quad n = 1, 2, 3, 4.$$

This yields the four phases $\pi/4$, $3\pi/4$, $5\pi/4$ and $7\pi/4$ as needed.

This results in a two-dimensional signal space with unit basis functions

$$\phi_1(t) = \sqrt{\frac{2}{T_s}} \cos\left(2\pi f_c t\right)$$

$$\phi_2(t) = \sqrt{\frac{2}{T_s}} \sin\left(2\pi f_c t\right)$$

The first basis function is used as the in-phase component of the signal and the second as the quadrature component of the signal.

Hence, the signal constellation consists of the signal-space 4 points

$$\left(\pm\sqrt{\frac{E_s}{2}}, \pm\sqrt{\frac{E_s}{2}} \right).$$

The factors of 1/2 indicate that the total power is split equally between the two carriers.

Comparing these basis functions with that for BPSK shows clearly how QPSK can be viewed as two independent BPSK signals. Note that the signal-space points for BPSK do not need to split the symbol (bit) energy over the two carriers in the scheme shown in the BPSK constellation diagram.

QPSK systems can be implemented in a number of ways. An illustration of the major components of the transmitter and receiver structure are shown below.

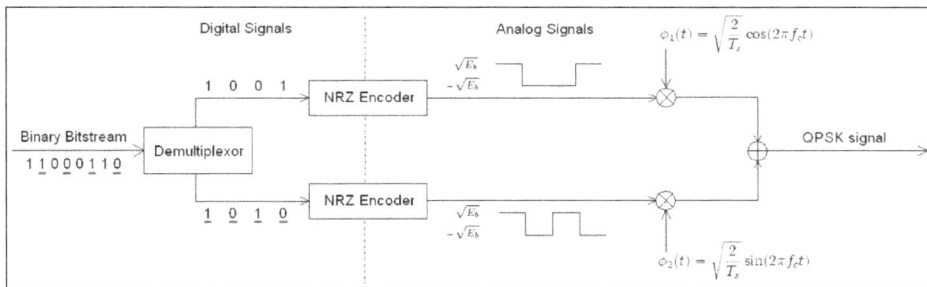

Conceptual transmitter structure for QPSK. The binary data stream is split into the in-phase and quadrature-phase components. These are then separately modulated onto two orthogonal basis functions. In this implementation, two sinusoids are used. Afterwards, the two signals are superimposed, and the resulting signal is the QPSK signal. Note the use of polar non-return-to-zero encoding. These encoders can be placed before for binary data source, but have been placed after to illustrate the conceptual difference between digital and analog signals involved with digital modulation.

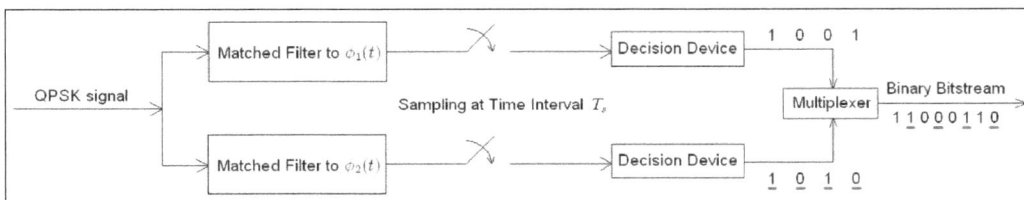

Receiver structure for QPSK. The matched filters can be replaced with correlators. Each detection device uses a reference threshold value to determine whether a 1 or 0 is detected.

Bit Error Rate

Although QPSK can be viewed as a quaternary modulation, it is easier to see it as two independent-ly modulated quadrature carriers. With this interpretation, the even (or odd) bits are used to mod-ulate the in-phase component of the carrier, while the odd (or even) bits are used to modulate the quadrature-phase component of the carrier. BPSK is used on both carriers and they can be inde-pendently demodulated.

As a result, the probability of bit-error for QPSK is the same as for BPSK:

$$P_b = Q\left(\sqrt{\frac{2E_b}{N_0}}\right)$$

However, in order to achieve the same bit-error probability as BPSK, QPSK uses twice the power (since two bits are transmitted simultaneously).

The symbol error rate is given by:

$$P_s = 1 - \left(1 - P_b\right)^2$$

$$= 2Q\left(\sqrt{\frac{E_s}{N_0}}\right) - \left[Q\left(\sqrt{\frac{E_s}{N_0}}\right)\right]^2$$

If the signal-to-noise ratio is high (as is necessary for practical QPSK systems) the probability of symbol error may be approximated:

$$P_s \approx 2Q\left(\sqrt{\frac{E_s}{N_0}}\right) = \text{erfc}\left(\sqrt{\frac{E_s}{2N_0}}\right) = \text{erfc}\left(\sqrt{\frac{E_b}{N_0}}\right)$$

The modulated signal is shown below for a short segment of a random binary data-stream. The two carrier waves are a cosine wave and a sine wave, as indicated by the signal-space analysis above. Here, the odd-numbered bits have been assigned to the in-phase component and the even-num-bered bits to the quadrature component (taking the first bit as number 1). The total signal – the sum of the two components – is shown at the bottom. Jumps in phase can be seen as the PSK changes the phase on each component at the start of each bit-period.

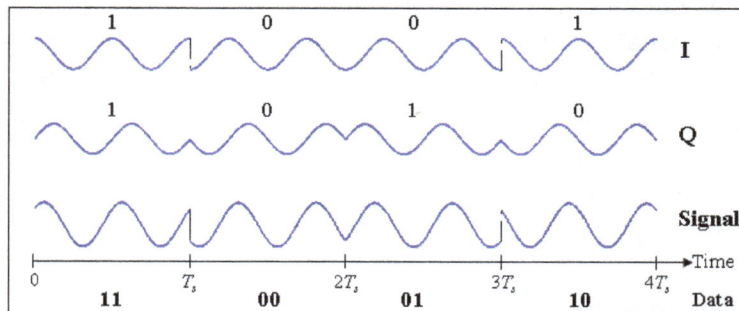

Timing diagram for QPSK. The binary data stream is shown beneath the time axis. The two signal components with their bit assignments are shown at the top, and the total combined signal at the bottom. Note the abrupt changes in phase at some of the bit-period boundaries.

The binary data that is conveyed by this waveform is: 11000110.

- The odd bits, highlighted here, contribute to the in-phase component: 11000110.

- The even bits, highlighted here, contribute to the quadrature-phase component: 11000110.

Variants

Offset QPSK (OQPSK)

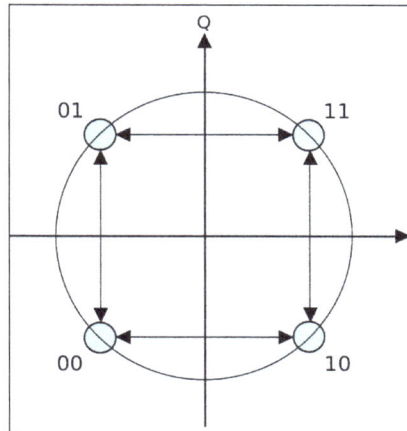

Signal doesn't pass through the origin, because only one bit of the symbol is changed at a time.

Offset quadrature phase-shift keying (OQPSK) is a variant of phase-shift keying modulation using four different values of the phase to transmit. It is sometimes called staggered quadrature phase-shift keying (SQPSK).

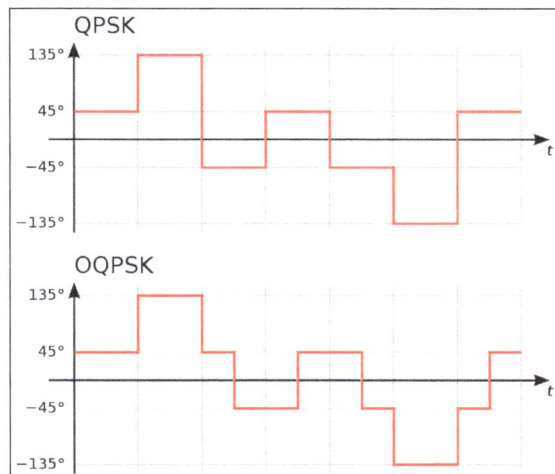

Difference of the phase between QPSK and OQPSK.

Taking four values of the phase (two bits) at a time to construct a QPSK symbol can allow the phase of the signal to jump by as much as 180° at a time. When the signal is low-pass filtered (as is typical in a transmitter), these phase-shifts result in large amplitude fluctuations, an undesirable quality in communication systems. By offsetting the timing of the odd and even bits by one bit-period, or half a symbol-period, the in-phase and quadrature components will never change at the same time. In the constellation diagram shown on the right, it can be seen that this will limit the phase-shift to no more than 90° at a time. This yields much lower amplitude fluctuations than non-offset QPSK and is sometimes preferred in practice.

The image above shows the difference in the behavior of the phase between ordinary QPSK and OQPSK. It can be seen that in the first plot the phase can change by 180° at once, while in OQPSK the changes are never greater than 90°.

The modulated signal is for a short segment of a random binary data-stream. Note the half symbol-period offset between the two component waves. The sudden phase-shifts occur about twice as often as for QPSK (since the signals no longer change together), but they are less severe. In other words, the magnitude of jumps is smaller in OQPSK when compared to QPSK.

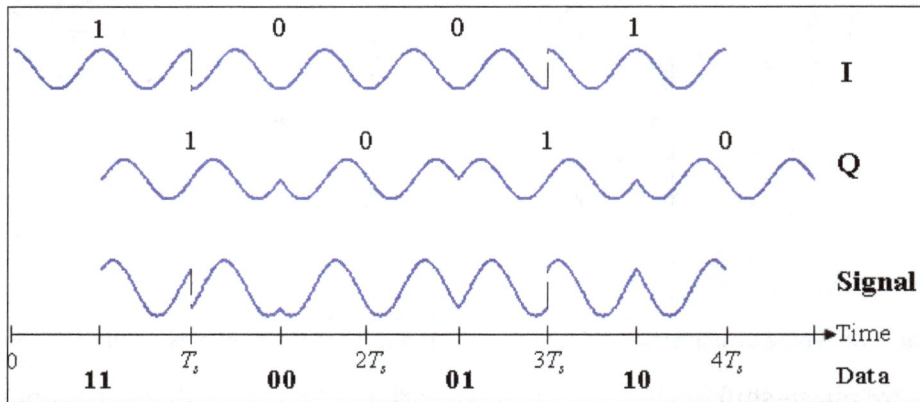

Timing diagram for offset-QPSK. The binary data stream is shown beneath the time axis. The two signal components with their bit assignments are shown the top and the total, combined signal at the bottom. Note the half-period offset between the two signal components.

π/4-QPSK

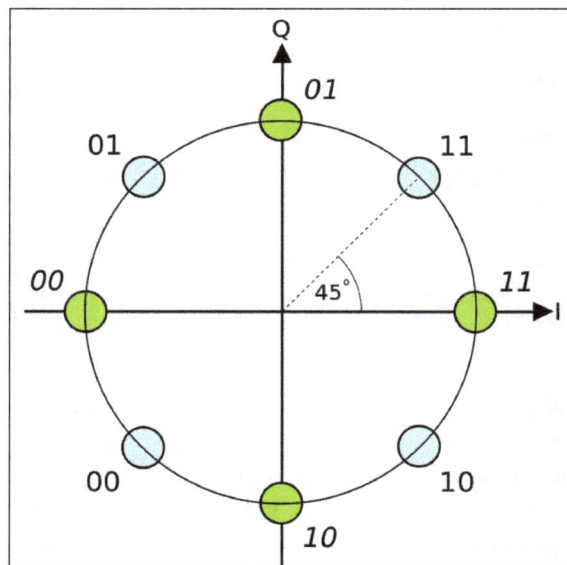

Dual constellation diagram for π/4-QPSK. This shows the two separate constellations with identical Gray coding but rotated by 45° with respect to each other.

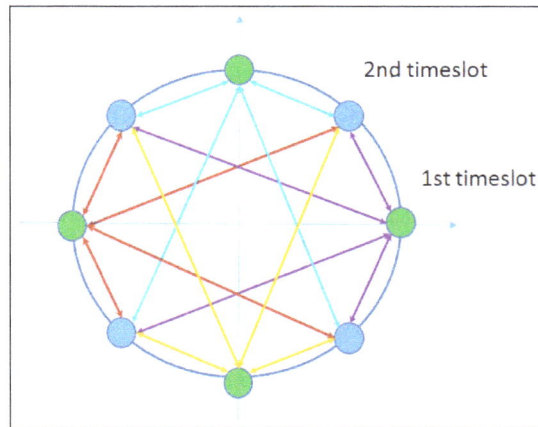

Transition scheme of the modulation symbols of the
π/4-QPSK (signal constellation). No zero crossings.

This variant of QPSK uses two identical constellations which are rotated by 45° ($\pi/4$ radians, hence the name) with respect to one another. Usually, either the even or odd symbols are used to select points from one of the constellations and the other symbols select points from the other constellation. This also reduces the phase-shifts from a maximum of 180°, but only to a maximum of 135° and so the amplitude fluctuations of $\pi/4$-QPSK are between OQPSK and non-offset QPSK.

One property this modulation scheme possesses is that if the modulated signal is represented in the complex domain, transitions between symbols never pass through 0. In other words, the signal does not pass through the origin. This lowers the dynamical range of fluctuations in the signal which is desirable when engineering communications signals.

On the other hand, $\pi/4$-QPSK lends itself to easy demodulation and has been adopted for use in, for example, TDMA cellular telephone systems.

The modulated signal is shown below for a short segment of a random binary data-stream. The construction is the same as above for ordinary QPSK. Successive symbols are taken from the two constellations shown in the diagram. Thus, the first symbol (1 1) is taken from the "blue" constellation and the second symbol (0 0) is taken from the "green" constellation. Note that magnitudes of the two component waves change as they switch between constellations, but the total signal's magnitude remains constant (constant envelope). The phase-shifts are between those of the two previous timing-diagrams.

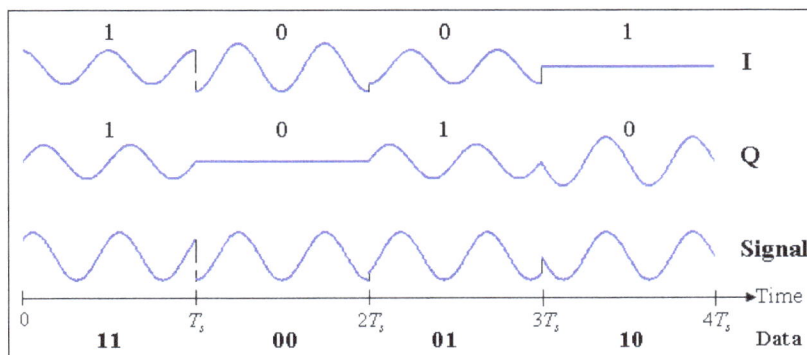

Timing diagram for π/4-QPSK. The binary data stream is shown beneath the time axis. The two signal components with their bit assignments are shown the top and the total, combined signal at the bottom.

SOQPSK

The license-free shaped-offset QPSK (SOQPSK) is interoperable with Feher-patented QPSK (FQPSK), in the sense that an integrate-and-dump offset QPSK detector produces the same output no matter which kind of transmitter is used.

These modulations carefully shape the I and Q waveforms such that they change very smoothly, and the signal stays constant-amplitude even during signal transitions. (Rather than traveling instantly from one symbol to another, or even linearly, it travels smoothly around the constant-amplitude circle from one symbol to the next.)

The standard description of SOQPSK-TG involves ternary symbols.

DPQPSK

Dual-polarization quadrature phase shift keying (DPQPSK) or dual-polarization QPSK - involves the polarization multiplexing of two different QPSK signals, thus improving the spectral efficiency by a factor of 2. This is a cost-effective alternative to utilizing 16-PSK, instead of QPSK to double the spectral efficiency.

Higher-order PSK

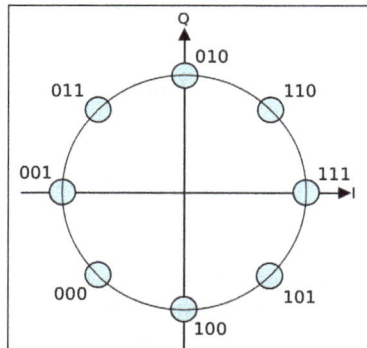

Constellation diagram for 8-PSK with Gray coding.

Any number of phases may be used to construct a PSK constellation but 8-PSK is usually the highest order PSK constellation deployed. With more than 8 phases, the error-rate becomes too high and there are better, though more complex, modulations available such as quadrature amplitude modulation (QAM). Although any number of phases may be used, the fact that the constellation must usually deal with binary data means that the number of symbols is usually a power of 2 to allow an integer number of bits per symbol.

Bit Error Rate

For the general M-PSK there is no simple expression for the symbol-error probability if $M > 4$. Unfortunately, it can only be obtained from,

$$P_s = 1 - \int_{-\pi/M}^{\pi/M} p_{\theta_r}(\theta_r) d\theta_r,$$

where,

$$p_{\theta_r}(\theta_r) = \frac{1}{2\pi} e^{-2\gamma_s \sin^2 \theta_r} \int_0^\infty V e^{-\frac{1}{2}\left(V - 2\sqrt{\gamma_s} \cos \theta_r\right)^2} \, dV,$$

$$V = \sqrt{r_1^2 + r_2^2},$$

$$\theta_r = \tan^{-1}\left(\frac{r_2}{r_1}\right),$$

$$\gamma_s = \frac{E_s}{N_0}$$

and

$$r_1 \sim N\left(\sqrt{E_s}, \frac{1}{2} N_0\right)$$

and $r_2 \sim N\left(0, \frac{1}{2} N_0\right)$ are jointly Gaussian random variables.

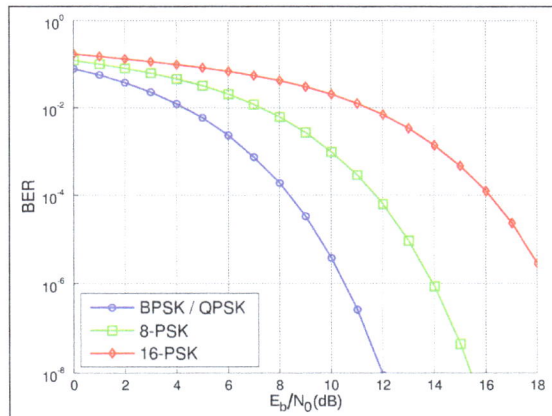

Bit-error rate curves for BPSK, QPSK, 8-PSK and
16-PSK, additive white Gaussian noise channel.

This may be approximated for high M and high E_b / N_0 by:

$$P_s \approx 2Q\left(\sqrt{2\gamma_s} \sin \frac{\pi}{M}\right).$$

The bit-error probability for M --PSK can only be determined exactly once the bit-mapping is known. However, when Gray coding is used, the most probable error from one symbol to the next produces only a single bit-error,

$$P_b \approx \frac{1}{k} P_s.$$

(Using Gray coding allows us to approximate the Lee distance of the errors as the Hamming distance of the errors in the decoded bitstream, which is easier to implement in hardware.)

The graph on the left compares the bit-error rates of BPSK, QPSK (which are the same, as noted above), 8-PSK and 16-PSK. It is seen that higher-order modulations exhibit higher error-rates; in exchange however they deliver a higher raw data-rate.

Bounds on the error rates of various digital modulation schemes can be computed with application of the union bound to the signal constellation.

Differential Phase-shift Keying (DPSK)

Differential Encoding

Differential phase shift keying (DPSK) is a common form of phase modulation that conveys data by changing the phase of the carrier wave. As mentioned for BPSK and QPSK there is an ambiguity of phase if the constellation is rotated by some effect in the communications channel through which the signal passes. This problem can be overcome by using the data to *change* rather than *set* the phase.

For example, in differentially encoded BPSK a binary "1" may be transmitted by adding 180° to the current phase and a binary "0" by adding 0° to the current phase. Another variant of DPSK is Symmetric Differential Phase Shift keying, SDPSK, where encoding would be +90° for a "1" and −90° for a "0".

In differentially encoded QPSK (DQPSK), the phase-shifts are 0°, 90°, 180°, −90° corresponding to data "00", "01", "11", "10". This kind of encoding may be demodulated in the same way as for non-differential PSK but the phase ambiguities can be ignored. Thus, each received symbol is demodulated to one of the points in the constellation and a comparator then computes the difference in phase between this received signal and the preceding one. The difference encodes the data as described above. Symmetric Differential Quadrature Phase Shift Keying (SDQPSK) is like DQPSK, but encoding is symmetric, using phase shift values of −135°, −45°, +45° and +135°.

The modulated signal is shown below for both DBPSK and DQPSK as described above. In the figure, it is assumed that the *signal starts with zero phase*, and so there is a phase shift in both signals at $t = 0$.

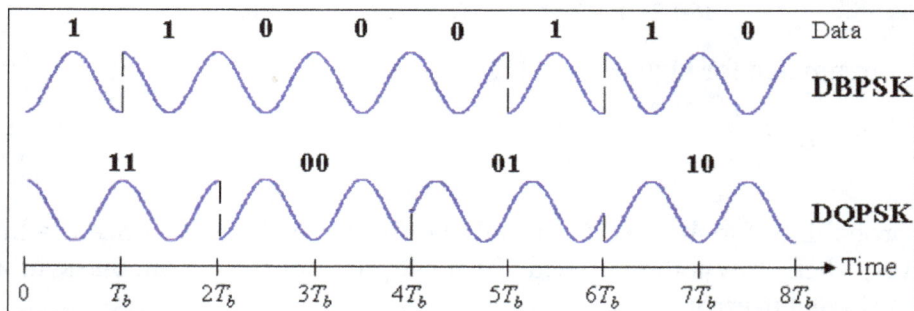

Timing diagram for DBPSK and DQPSK. The binary data stream is above the DBPSK signal. The individual bits of the DBPSK signal are grouped into pairs for the DQPSK signal, which only changes every $T_s = 2T_b$.

Analysis shows that differential encoding approximately doubles the error rate compared to ordinary $M - \text{PSK}$ but this may be overcome by only a small increase in E_b / N_0. Furthermore, this analysis (and the graphical results below) are based on a system in which the only corruption is additive white Gaussian noise (AWGN). However, there will also be a physical channel between the transmitter and receiver in the communication system. This channel will, in general, introduce an unknown phase-shift to the PSK signal; in these cases the differential schemes can yield a *better* error-rate than the ordinary schemes which rely on precise phase information.

One of the most popular applications of DPSK is the Bluetooth standard where $\pi/4$-DQPSK and 8-DPSK were implemented.

Demodulation

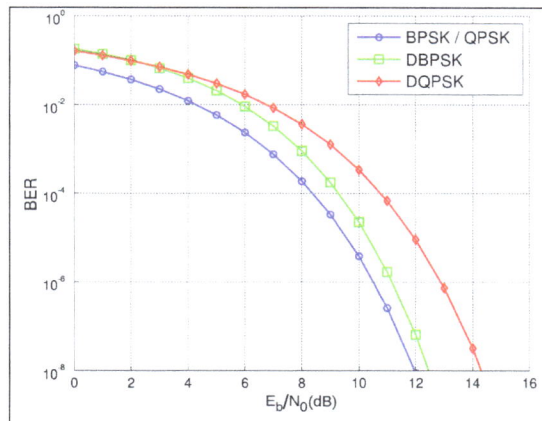

BER comparison between DBPSK, DQPSK and their non-differential
forms using Gray coding and operating in white noise

For a signal that has been differentially encoded, there is an obvious alternative method of demodulation. Instead of demodulating as usual and ignoring carrier-phase ambiguity, the phase between two successive received symbols is compared and used to determine what the data must have been. When differential encoding is used in this manner, the scheme is known as differential phase-shift keying (DPSK). Note that this is subtly different from just differentially encoded PSK since, upon reception, the received symbols are *not* decoded one-by-one to constellation points but are instead compared directly to one another.

Call the received symbol in the kth timeslot r_k and let it have phase ϕ_k. Assume without loss of generality that the phase of the carrier wave is zero. Denote the additive white Gaussian noise (AWGN) term as n_k. Then,

$$r_k = \sqrt{E_s} e^{j\phi_k} + n_k.$$

The decision variable for the $k-1^{th}$ symbol and the kth symbol is the phase difference between r_k and r_{k-1}. That is, if r_k is projected onto r_{k-1}, the decision is taken on the phase of the resultant complex number,

$$r_k r_{k-1}^* = E_s e^{j(\varphi_k - \varphi_{k-1})} + \sqrt{E_s} e^{j\varphi_k} n_{k-1}^* + \sqrt{E_s} e^{-j\varphi_{k-1}} n_k + n_k n_{k-1}^*$$

where superscript * denotes complex conjugation. In the absence of noise, the phase of this is $\phi_k - \phi_{k-1}$, the phase-shift between the two received signals which can be used to determine the data transmitted.

The probability of error for DPSK is difficult to calculate in general, but, in the case of DBPSK it is,

$$P_b = \frac{1}{2} e^{-\frac{E_b}{N_0}},$$

which, when numerically evaluated, is only slightly worse than ordinary BPSK, particularly at higher E_b / N_0 values.

Using DPSK avoids the need for possibly complex carrier-recovery schemes to provide an accurate phase estimate and can be an attractive alternative to ordinary PSK.

In optical communications, the data can be modulated onto the phase of a laser in a differential way. The modulation is a laser which emits a continuous wave, and a Mach–Zehnder modulator which receives electrical binary data. For the case of BPSK, the laser transmits the field unchanged for binary '1', and with reverse polarity for '0'. The demodulator consists of a delay line interferometer which delays one bit, so two bits can be compared at one time. In further processing, a photodiode is used to transform the optical field into an electric current, so the information is changed back into its original state.

The bit-error rates of DBPSK and DQPSK are compared to their non-differential counterparts in the graph to the right. The loss for using DBPSK is small enough compared to the complexity reduction that it is often used in communications systems that would otherwise use BPSK. For DQPSK though, the loss in performance compared to ordinary QPSK is larger and the system designer must balance this against the reduction in complexity.

Example - Differentially encoded BPSK:

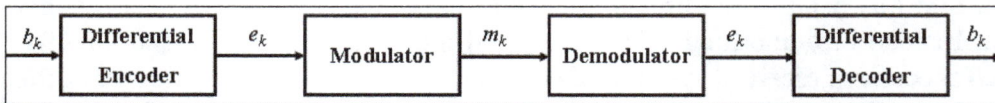

Differential encoding/decoding system diagram.

At the kth time-slot call the bit to be modulated b_k, the differentially encoded bit e_k and the resulting modulated signal $m_k(t)$. Assume that the constellation diagram positions the symbols at ±1 (which is BPSK). The differential encoder produces where $\oplus$ indicates binary or modulo-2 addition.

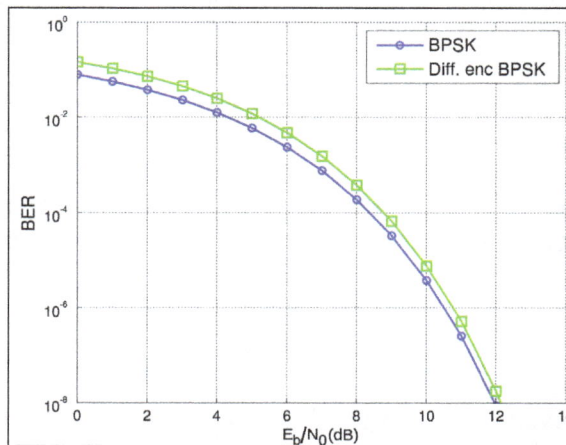

BER comparison between BPSK and differentially encoded
BPSK with Gray coding operating in white noise.

So e_k only changes state (from binary "0" to binary "1" or from binary "1" to binary "0") if is a binary "1". Otherwise it remains in its previous state. This is the description of differentially encoded BPSK given above.

The received signal is demodulated to yield $e_k = \pm 1$ and then the differential decoder reverses the encoding procedure and produces,

$$b_k = e_k \oplus e_{k-1},$$

since binary subtraction is the same as binary addition.

Therefore, $b_k = 1$ if e_k and e_{k-1} differ and $b_k = 0$ if they are the same. Hence, if both e_k and e_{k-1} are *inverted*, b_k will still be decoded correctly. Thus, the 180° phase ambiguity does not matter.

Differential schemes for other PSK modulations may be devised along similar lines. The waveforms for DPSK are the same as for differentially encoded PSK given above since the only change between the two schemes is at the receiver.

The BER curve for this example is compared to ordinary BPSK on the right. As mentioned above, whilst the error rate is approximately doubled, the increase needed in E_s / N_0 to overcome this is small. The increase in E_s / N_0 required to overcome differential modulation in coded systems, however, is larger – typically about 3 dB. The performance degradation is a result of noncoherent transmission – in this case it refers to the fact that tracking of the phase is completely ignored.

Mutual information with Additive white Gaussian Noise

The mutual information of PSK can be evaluated in additive Gaussian noise by numerical integration of its definition. The curves of mutual information saturate to the number of bits carried by each symbol in the limit of infinite signal to noise ratio E_s / N_0. On the contrary, in the limit of small signal to noise ratios the mutual information approaches the AWGN channel capacity, which is the supremum among all possible choices of symbol statistical distributions.

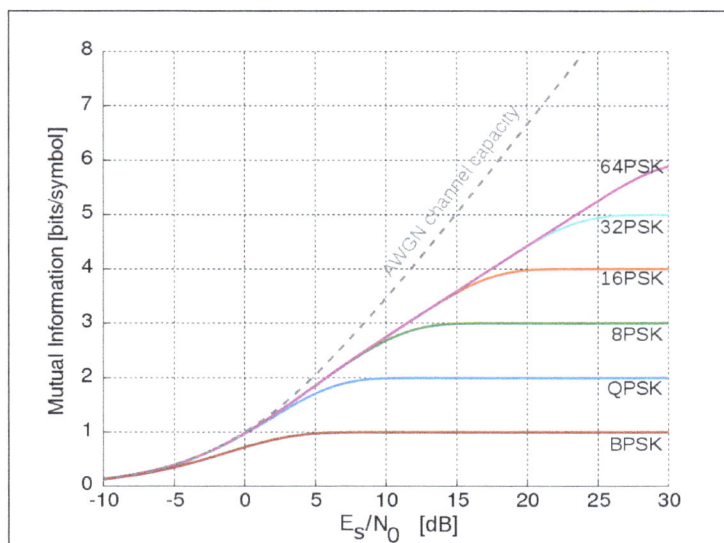

Mutual information of PSK over the AWGN channel.

At intermediate values of signal to noise ratios the mutual information (MI) is well approximated by:

$$\text{MI} \simeq \log_2 \left(\sqrt{\frac{4\pi}{e} \frac{E_s}{N_0}} \right).$$

The mutual information of PSK over the AWGN channel is generally farther to the AWGN channel capacity than QAM modulation formats.

M-ARY ENCODING

The word binary represents two bits. M represents a digit that corresponds to the number of conditions, levels, or combinations possible for a given number of binary variables.

This is the type of digital modulation technique used for data transmission in which instead of one bit, two or more bits are transmitted at a time. As a single signal is used for multiple bit transmission, the channel bandwidth is reduced.

M-ary Equation

If a digital signal is given under four conditions, such as voltage levels, frequencies, phases, and amplitude, then M = 4.

The number of bits necessary to produce a given number of conditions is expressed mathematically as,

$$N = log_2 M$$

Where,

N is the number of bits necessary.

M is the number of conditions, levels, or combinations possible with N bits.

The above equation can be re-arranged as,

$$2^N = M$$

For example, with two bits, $2^2 = 4$ conditions are possible.

Types of M-ary Techniques

In general, Multi-level (M-ary) modulation techniques are used in digital communications as the digital inputs with more than two modulation levels are allowed on the transmitter's input. Hence, these techniques are bandwidth efficient.

There are many M-ary modulation techniques. Some of these techniques, modulate one parameter of the carrier signal, such as amplitude, phase, and frequency.

M-ary ASK

This is called M-ary Amplitude Shift Keying (M-ASK) or M-ary Pulse Amplitude Modulation (PAM).

The amplitude of the carrier signal, takes on M different levels.

Representation of M-ary ASK

$$S_m(t) = A_m cos(2\pi f_c t), \ Am\epsilon(2m-1-M)\Delta, m=1,2....M \text{ and } 0 \le t \le T_s.$$

Some prominent features of M-ary ASK are:

- This method is also used in PAM.

- Its implementation is simple.

- M-ary ASK is susceptible to noise and distortion.

M-ary FSK

This is called as M-ary Frequency Shift Keying (M-ary FSK). The frequency of the carrier signal, takes on M different levels.

Representation of M-ary FSK

$$S_i(t) = \sqrt{\frac{2E_s}{T_s}} cos(\frac{\pi}{T_s}(n_c + i)t)$$

$$0 \le t \le T_s \text{ and } i=1,2,3.....M$$

Where $f_c = \frac{nc}{2T_s}$ for some fixed integer n.

Some prominent features of M-ary FSK are:

- Not susceptible to noise as much as ASK.

- The transmitted M number of signals are equal in energy and duration.

- The signals are separated by $\frac{1}{2T_s}$ Hz making the signals orthogonal to each other.

- Since M signals are orthogonal, there is no crowding in the signal space.

- The bandwidth efficiency of M-ary FSK decreases and the power efficiency increases with the increase in M.

M-ary PSK

This is called as M-ary Phase Shift Keying (M-ary PSK). The phase of the carrier signal, takes on M different levels.

Representation of M-ary PSK

$$S_i(t) = \sqrt{\frac{2E}{T}} cos(w_o t + \phi_i t), \ 0 \le t \le T \text{ and } i=1, 2...M$$

$$\phi_i(t) = \frac{2\pi i}{M}, \text{ where } i = 1, 2, 3......M$$

Some prominent features of M-ary PSK are:

- The envelope is constant with more phase possibilities.

- This method was used during the early days of space communication.

- Better performance than ASK and FSK.

- Minimal phase estimation error at the receiver.

- The bandwidth efficiency of M-ary PSK decreases and the power efficiency increases with the increase in M.

So far, we have discussed different modulation techniques. The output of all these techniques is a binary sequence, represented as 1s and 0s.

4

Multiple Access Techniques

The techniques that let multiple mobile users share the allotted spectrum in the most effective manner for improving the overall access capacity are known as multiple access techniques. A few of such techniques are frequency division multiple access, time division multiple access and space division multiple access. The topics elaborated in this chapter will help in gaining a better perspective about these multiple access techniques.

Multiple access schemes are used to allow many mobile users to share simultaneously a finite amount of radio spectrum.

In wireless communication systems, it is often desirable to allow the subscriber to send information simultaneously from the mobile station to the base station while receiving information from the base station to the mobile station.

A cellular system divides any given area into cells where a mobile unit in each cell communicates with a base station. The main aim in the cellular system design is to be able to increase the capacity of the channel, i.e., to handle as many calls as possible in a given bandwidth with a sufficient level of quality of service.

There are several different ways to allow access to the channel. These includes mainly the following:

- Requency division multiple-access (FDMA).

- Time division multiple-access (TDMA).

- Code division multiple-access (CDMA).

- Space division multiple access (SDMA).

Depending on how the available bandwidth is allocated to the users, these techniques can be classified as narrowband and wideband systems.

Narrowband Systems

Systems operating with channels substantially narrower than the coherence bandwidth are called as Narrow band systems. Narrow band TDMA allows users to use the same channel but allocates a unique time slot to each user on the channel, thus separating a small number of users in time on a single channel.

Wideband Systems

In wideband systems, the transmission bandwidth of a single channel is much larger than the coherence bandwidth of the channel. Thus, multipath fading doesn't greatly affect the received signal within a wideband channel, and frequency selective fades occur only in a small fraction of the signal bandwidth.

Frequency Division Multiple Access (FDMA)

FDMA is the basic technology for advanced mobile phone services. The features of FDMA are as follows.

- FDMA allots a different sub-band of frequency to each different user to access the network.

- If FDMA is not in use, the channel is left idle instead of allotting to the other users.

- FDMA is implemented in Narrowband systems and it is less complex than TDMA.

- Tight filtering is done here to reduce adjacent channel interference.

- The base station BS and mobile station MS, transmit and receive simultaneously and continuously in FDMA.

Time Division Multiple Access (TDMA)

In the cases where continuous transmission is not required, there TDMA is used instead of FDMA. The features of TDMA include the following.

- TDMA shares a single carrier frequency with several users where each users makes use of non-overlapping time slots.

- Data transmission in TDMA is not continuous, but occurs in bursts. Hence handsoff process is simpler.

- TDMA uses different time slots for transmission and reception thus duplexers are not required.

- TDMA has an advantage that is possible to allocate different numbers of time slots per frame to different users.

- Bandwidth can be supplied on demand to different users by concatenating or reassigning time slot based on priority.

Code Division Multiple Access (CDMA)

Code division multiple access technique is an example of multiple access where several transmitters use a single channel to send information simultaneously. Its features are as follows.

- In CDMA every user uses the full available spectrum instead of getting allotted by separate frequency.

- CDMA is much recommended for voice and data communications.

- While multiple codes occupy the same channel in CDMA, the users having same code can communicate with each other.

- CDMA offers more air-space capacity than TDMA.

- The hands-off between base stations is very well handled by CDMA.

Space Division Multiple Access (SDMA)

Space division multiple access or spatial division multiple access is a technique which is MIMO (multiple-input multiple-output) architecture and used mostly in wireless and satellite communication. It has the following features.

- All users can communicate at the same time using the same channel.

- SDMA is completely free from interference.

- A single satellite can communicate with more satellites receivers of the same frequency.

- The directional spot-beam antennas are used and hence the base station in SDMA, can track a moving user.

- Controls the radiated energy for each user in space.

Spread Spectrum Multiple Access

Spread spectrum multiple access (SSMA) uses signals which have a transmission bandwidth whose magnitude is greater than the minimum required RF bandwidth.

There are two main types of spread spectrum multiple access techniques –

- Frequency hopped spread spectrum (FHSS)

- Direct sequence spread spectrum (DSSS)

Frequency Hopped Spread Spectrum (FHSS)

This is a digital multiple access system in which the carrier frequencies of the individual users are varied in a pseudo random fashion within a wideband channel. The digital data is broken into uniform sized bursts which is then transmitted on different carrier frequencies.

Direct Sequence Spread Spectrum (DSSS)

This is the most commonly used technology for CDMA. In DS-SS, the message signal is multiplied by a Pseudo Random Noise Code. Each user is given his own code word which is orthogonal to the codes of other users and in order to detect the user, the receiver must know the code word used by the transmitter.

The combinational sequences called as hybrid are also used as another type of spread spectrum. Time hopping is also another type which is rarely mentioned.

Since many users can share the same spread spectrum bandwidth without interfering with one another, spread spectrum systems become bandwidth efficient in a multiple user environment.

FREQUENCY DIVISION MULTIPLE ACCESS

Frequency division multiple access (FDMA) is a channel access method used in some multiple-access protocols. FDMA allows multiple users to send data through a single communication channel, such as a coaxial cable or microwave beam, by dividing the bandwidth of the channel into separate non-overlapping frequency sub-channels and allocating each sub-channel to a separate user. Users can send data through a subchannel by modulating it on a carrier wave at the subchannel's frequency. It is used in satellite communication systems and telephone trunklines.

Method

Alternatives include time-division multiple access (TDMA), code-division multiple access (CDMA), or space-division multiple access (SDMA). These protocols are utilized differently, at different levels of the theoretical OSI model.

Disadvantage: Crosstalk may cause interference among frequencies and disrupt the transmission.

- In FDMA, all users share the satellite transponder or frequency channel simultaneously but each user transmits at single frequency.

- FDMA can be used with both analog and digital signal but it generally used with analog signal.

- FDMA requires high-performing filters in the radio hardware, in contrast to TDMA and CDMA.

- FDMA is not vulnerable to the timing problems that TDMA has. Since a predetermined frequency band is available for the entire period of communication, stream data (a continuous flow of data that may not be packetized) can easily be used with FDMA.

- Due to the frequency filtering, FDMA is not sensitive to near-far problem which is pronounced for CDMA.

- Each user transmits and receives at different frequencies as each user gets a unique frequency slots.

FDMA is distinct from frequency division duplexing (FDD). While FDMA allows multiple users simultaneous access to a transmission system, FDD refers to how the radio channel is shared between the uplink and downlink (for instance, the traffic going back and forth between a mobile-phone and a mobile phone base station). Frequency-division multiplexing (FDM) is also distinct from FDMA. FDM is a physical layer technique that combines and transmits low-bandwidth channels through a high-bandwidth channel. FDMA, on the other hand, is an access method in the data link layer.

FDMA also supports demand assignment in addition to fixed assignment. *Demand assignment* allows all users apparently continuous access of the radio spectrum by assigning carrier frequencies on a temporary basis using a statistical assignment process. The first FDMA *demand-assignment* system for satellite was developed by COMSAT for use on the *Intelsat* series *IVA* and *V* satellites.

There are two main techniques:

- Multi-channel per-carrier (MCPC).

- Single-channel per-carrier (SCPC).

TIME DIVISION MULTIPLE ACCESS

Time-division multiple access (TDMA) is a channel access method for shared-medium networks. It allows several users to share the same frequency channel by dividing the signal into different time slots. The users transmit in rapid succession, one after the other, each using its own time slot. This allows multiple stations to share the same transmission medium (e.g. radio frequency channel) while using only a part of its channel capacity. TDMA is used in the digital 2G cellular systems such as Global System for Mobile Communications (GSM), IS-136, Personal Digital Cellular (PDC) and iDEN, and in the Digital Enhanced Cordless Telecommunications (DECT) standard for portable phones. TDMA was first used in satellite communication systems by Western Union in its Westar 3 communications satellite in 1979. It is now used extensively in satellite communications, combat-net radio systems, and passive optical network (PON) networks for upstream traffic from premises to the operator.

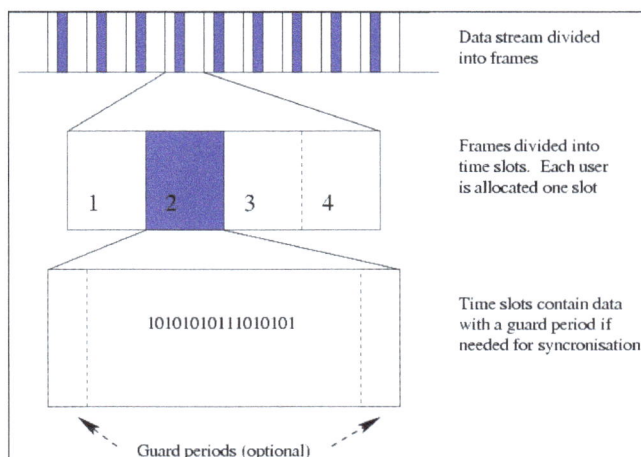

TDMA frame structure showing a data stream divided into frames and those frames divided into time slots.

TDMA is a type of time-division multiplexing (TDM), with the special point that instead of having one transmitter connected to one receiver, there are multiple transmitters. In the case of the *uplink* from a mobile phone to a base station this becomes particularly difficult because the mobile phone can move around and vary the *timing advance* required to make its transmission match the gap in transmission from its peers.

TDMA Characteristics

- Shares single carrier frequency with multiple users.

- Non-continuous transmission makes handoff simpler.

- Slots can be assigned on demand in dynamic TDMA.

- Less stringent power control than CDMA due to reduced intra cell interference.

- Higher synchronization overhead than CDMA.

- Advanced equalization may be necessary for high data rates if the channel is "frequency selective" and creates Intersymbol interference.

- Cell breathing (borrowing resources from adjacent cells) is more complicated than in CDMA.

- Frequency/slot allocation complexity.

- Pulsating power envelope: interference with other devices.

TDMA in Mobile Phone Systems

2G systems

Most 2G cellular systems, with the notable exception of IS-95, are based on TDMA. GSM, D-AMPS, PDC, iDEN, and PHS are examples of TDMA cellular systems. GSM combines TDMA with Frequency Hopping and wideband transmission to minimize common types of interference.

In the GSM system, the synchronization of the mobile phones is achieved by sending timing advance commands from the base station which instructs the mobile phone to transmit earlier and by how much. This compensates for the propagation delay resulting from the light speed velocity of radio waves. The mobile phone is not allowed to transmit for its entire time slot, but there is a guard interval at the end of each time slot. As the transmission moves into the guard period, the mobile network adjusts the timing advance to synchronize the transmission.

Initial synchronization of a phone requires even more care. Before a mobile transmits there is no way to actually know the offset required. For this reason, an entire time slot has to be dedicated to mobiles attempting to contact the network; this is known as the random-access channel (RACH) in GSM. The mobile attempts to broadcast at the beginning of the time slot, as received from the network. If the mobile is located next to the base station, there will be no time delay and this will succeed. If, however, the mobile phone is at just less than 35 km from the base station, the time delay will mean the mobile's broadcast arrives at the very end of the time slot. In that case, the mobile will be instructed to broadcast its messages starting nearly a whole time slot earlier than would be expected otherwise. Finally, if the mobile is beyond the 35 km cell range in GSM, then the RACH will arrive in a neighbouring time slot and be ignored. It is this feature, rather than limitations of power, that limits the range of a GSM cell to 35 km when no special extension techniques are used. By changing the synchronization between the uplink and downlink at the base station, however, this limitation can be overcome.

3G systems

Although most major 3G systems are primarily based upon CDMA, time-division duplexing (TDD), packet scheduling (dynamic TDMA) and packet oriented multiple access schemes are available in 3G form, combined with CDMA to take advantage of the benefits of both technologies.

While the most popular form of the UMTS 3G system uses CDMA and frequency division duplexing (FDD) instead of TDMA, TDMA is combined with CDMA and time-division duplexing in two standard UMTS UTRA.

TDMA in Wired Networks

The ITU-T G.hn standard, which provides high-speed local area networking over existing home wiring (power lines, phone lines and coaxial cables) is based on a TDMA scheme. In G.hn, a "master" device allocates "Contention-Free Transmission Opportunities" (CFTXOP) to other "slave" devices in the network. Only one device can use a CFTXOP at a time, thus avoiding collisions. FlexRay protocol which is also a wired network used for safety-critical communication in modern cars, uses the TDMA method for data transmission control.

Comparison with other Multiple-access Schemes

In radio systems, TDMA is usually used alongside frequency-division multiple access (FDMA) and frequency division duplex (FDD); the combination is referred to as FDMA/TDMA/FDD. This is the case in both GSM and IS-136 for example. Exceptions to this include the DECT and Personal Handy-phone System (PHS) micro-cellular systems, UMTS-TDD UMTS variant, and China's TD-SCDMA, which use time-division duplexing, where different time slots are allocated for the base station and handsets on the same frequency.

A major advantage of TDMA is that the radio part of the mobile only needs to listen and broadcast for its own time slot. For the rest of the time, the mobile can carry out measurements on the network, detecting surrounding transmitters on different frequencies. This allows safe inter frequency handovers, something which is difficult in CDMA systems, not supported at all in IS-95 and supported through complex system additions in Universal Mobile Telecommunications System (UMTS). This in turn allows for co-existence of microcell layers with macrocell layers.

CDMA, by comparison, supports "soft hand-off" which allows a mobile phone to be in communication with up to 6 base stations simultaneously, a type of "same-frequency handover". The incoming packets are compared for quality, and the best one is selected. CDMA's "cell breathing" characteristic, where a terminal on the boundary of two congested cells will be unable to receive a clear signal, can often negate this advantage during peak periods.

A disadvantage of TDMA systems is that they create interference at a frequency which is directly connected to the time slot length. This is the buzz which can sometimes be heard if a TDMA phone is left next to a radio or speakers. Another disadvantage is that the "dead time" between time slots limits the potential bandwidth of a TDMA channel. These are implemented in part because of the difficulty in ensuring that different terminals transmit at exactly the times required. Handsets that are moving will need to constantly adjust their timings to ensure their transmission is received at precisely the right time, because as they move further from the base station, their signal will take longer to arrive. This also means that the major TDMA systems have hard limits on cell sizes in terms of range, though in practice the power levels required to receive and transmit over distances greater than the supported range would be mostly impractical anyway.

Dynamic TDMA

In dynamic time-division multiple access (dynamic TDMA), a scheduling algorithm dynamically reserves a variable number of time slots in each frame to variable bit-rate data streams, based on the traffic demand of each data stream. Dynamic TDMA is used in:

- HIPERLAN/2 broadband radio access network.

- IEEE 802.16a WiMax.

- Bluetooth.

- Military Radios/Tactical Data Link.

- TD-SCDMA.

- ITU-T G.hn.

- Simulation of TDMA/DTMA links.

CODE DIVISION MULTIPLE ACCESS

Code-division multiple access (CDMA) is a channel access method used by various radio communication technologies.

CDMA is an example of multiple access, where several transmitters can send information simultaneously over a single communication channel. This allows several users to share a band of frequencies. To permit this without undue interference between the users, CDMA employs spread spectrum technology and a special coding scheme (where each transmitter is assigned a code).

CDMA is used as the access method in many mobile phone standards. IS-95, also called "cdma-One", and its 3G evolution CDMA2000, are often simply referred to as "CDMA", but UMTS, the 3G standard used by GSM carriers, also uses "wideband CDMA", or W-CDMA, as well as TD-CDMA and TD-SCDMA, as its radio technologies.

Uses

- One of the early applications for code-division multiplexing is in the Global Positioning System (GPS). This predates and is distinct from its use in mobile phones.

- The Qualcomm standard IS-95, marketed as cdmaOne.

- The Qualcomm standard IS-2000, known as CDMA2000, is used by several mobile phone companies, including the Globalstar network.

- The UMTS 3G mobile phone standard, which uses W-CDMA.

- CDMA has been used in the OmniTRACS satellite system for transportation logistics.

A CDMA2000 mobile phone.

Steps in CDMA Modulation

CDMA is a spread-spectrum multiple-access technique. A spread-spectrum technique spreads the bandwidth of the data uniformly for the same transmitted power. A spreading code is a pseudo-random code that has a narrow ambiguity function, unlike other narrow pulse codes. In CDMA a locally generated code runs at a much higher rate than the data to be transmitted. Data for transmission is combined by bitwise XOR (exclusive OR) with the faster code. The figure shows how a spread-spectrum signal is generated. The data signal with pulse duration of T_b (symbol period) is XORed with the code signal with pulse T_c duration of T_c (chip period). (Note: bandwidth is proportional to $1/T$, where $T =$ bit time.) Therefore, the bandwidth of the data signal is $1/T_b$ and the bandwidth of the spread spectrum signal is $1/T_c$. Since T_c is much smaller than T_b, the bandwidth of the spread-spectrum signal is much larger than the bandwidth of the original signal. The ratio T_b/T_c is called the spreading factor or processing gain and determines to a certain extent the upper limit of the total number of users supported simultaneously by a base station.

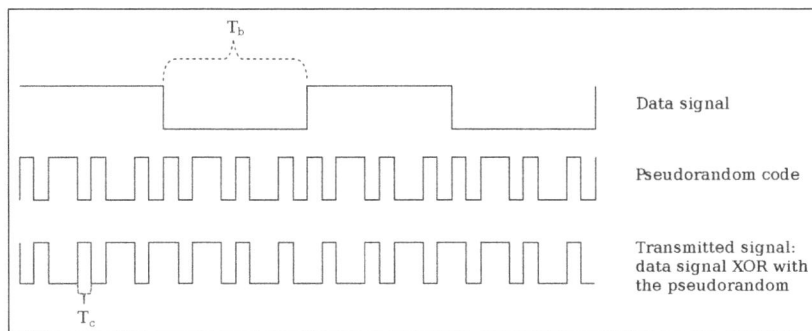

Generation of a CDMA signal.

Each user in a CDMA system uses a different code to modulate their signal. Choosing the codes used to modulate the signal is very important in the performance of CDMA systems. The best performance occurs when there is good separation between the signal of a desired user and the signals of other users. The separation of the signals is made by correlating the received signal with the locally generated code of the desired user. If the signal matches the desired user's code, then the correlation function will be high and the system can extract that signal. If the desired user's code has nothing in common with the signal, the correlation should be as close to zero as possible (thus eliminating the signal); this is referred to as cross-correlation. If the code is correlated with the signal at any time offset other than zero, the correlation should be as close to zero as possible. This is referred to as auto-correlation and is used to reject multi-path interference.

An analogy to the problem of multiple access is a room (channel) in which people wish to talk to each other simultaneously. To avoid confusion, people could take turns speaking (time division), speak at different pitches (frequency division), or speak in different languages (code division). CDMA is analogous to the last example where people speaking the same language can understand each other, but other languages are perceived as noise and rejected. Similarly, in radio CDMA, each group of users is given a shared code. Many codes occupy the same channel, but only users associated with a particular code can communicate.

In general, CDMA belongs to two basic categories: synchronous (orthogonal codes) and asynchronous (pseudorandom codes).

Code-division Multiplexing (Synchronous CDMA)

The digital modulation method is analogous to those used in simple radio transceivers. In the analog case, a low-frequency data signal is time-multiplied with a high-frequency pure sine-wave carrier and transmitted. This is effectively a frequency convolution (Wiener–Khinchin theorem) of the two signals, resulting in a carrier with narrow sidebands. In the digital case, the sinusoidal carrier is replaced by Walsh functions. These are binary square waves that form a complete orthonormal set. The data signal is also binary and the time multiplication is achieved with a simple XOR function. This is usually a Gilbert cell mixer in the circuitry.

Synchronous CDMA exploits mathematical properties of orthogonality between vectors representing the data strings. For example, binary string *1011* is represented by the vector (1, 0, 1, 1). Vectors can be multiplied by taking their dot product, by summing the products of their respective components (for example, if u = (a, b) and v = (c, d), then their dot product u·v = ac + bd). If the dot product is zero, the two vectors are said to be *orthogonal* to each other. Some properties of the dot product aid understanding of how W-CDMA works. If vectors a and b are orthogonal, then $\mathbf{a} \cdot \mathbf{b} = 0$ and:

$$a \cdot (a + b) = \| a \|^2 \text{, since } a \cdot a + a \cdot b = \| a \|^2 + 0,$$

$$a \cdot (-a + b) = -\| a \|^2 \text{, since } -a \cdot a + a \cdot b = -\| a \|^2 + 0,$$

$$b \cdot (a + b) = \| b \|^2 \text{, since } b \cdot a + b \cdot b = 0 + \| b \|^2,$$

$$b \cdot (a - b) = -\| b \|^2 \text{, since } b \cdot a - b \cdot b = 0 - \| b \|^2.$$

Each user in synchronous CDMA uses a code orthogonal to the others' codes to modulate their signal. An example of 4 mutually orthogonal digital signals is shown in the figure below. Orthogonal codes have a cross-correlation equal to zero; in other words, they do not interfere with each other. In the case of IS-95, 64-bit Walsh codes are used to encode the signal to separate different users. Since each of the 64 Walsh codes is orthogonal to all other, the signals are channelized into 64 orthogonal signals. The following example demonstrates how each user's signal can be encoded and decoded.

Example:

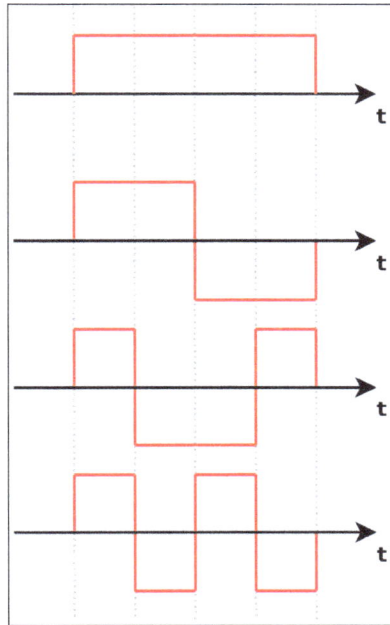

An example of 4 mutually orthogonal digital signals.

Start with a set of vectors that are mutually orthogonal. (Although mutual orthogonality is the only condition, these vectors are usually constructed for ease of decoding, for example columns or rows from Walsh matrices.) An example of orthogonal functions is shown in the adjacent picture. These vectors will be assigned to individual users and are called the *code*, *chip code*, or *chipping code*. In the interest of brevity, the rest of this example uses codes v with only two bits.

Each user is associated with a different code, say v. A 1 bit is represented by transmitting a positive code v, and a 0 bit is represented by a negative code −v. For example, if $v = (v_0, v_1) = (1, -1)$ and the data that the user wishes to transmit is $(1, 0, 1, 1)$, then the transmitted symbols would be,

$$(v, -v, v, v) = (v_0, v_1, -v_0, -v_1, v_0, v_1, v_0, v_1) = (1, -1, -1, 1, 1, -1, 1, -1).$$

We call this constructed vector the *transmitted vector*.

Each sender has a different, unique vector v chosen from that set, but the construction method of the transmitted vector is identical.

Now, due to physical properties of interference, if two signals at a point are in phase, they add to give twice the amplitude of each signal, but if they are out of phase, they subtract and give a signal that is the difference of the amplitudes. Digitally, this behaviour can be modelled by the addition of the transmission vectors, component by component.

If sender0 has code $(1, -1)$ and data $(1, 0, 1, 1)$, and sender1 has code $(1, 1)$ and data $(0, 0, 1, 1)$, and both senders transmit simultaneously, then this table describes the coding steps:

Step	Encode sender0	Encode sender1
0	code0 = $(1, -1)$, data0 = $(1, 0, 1, 1)$	code1 = $(1, 1)$, data1 = $(0, 0, 1, 1)$
1	encode0 = $2(1, 0, 1, 1) - (1, 1, 1, 1) = (1, -1, 1, 1)$	encode1 = $2(0, 0, 1, 1) - (1, 1, 1, 1) = (-1, -1, 1, 1)$
2	signal0 = encode0 $\otimes$ code0 $= (1, -1, 1, 1) \otimes (1, -1)$ $= (1, -1, -1, 1, 1, -1, 1, -1)$	signal1 = encode1 $\otimes$ code1 $= (-1, -1, 1, 1) \otimes (1, 1)$ $= (-1, -1, -1, -1, 1, 1, 1, 1)$

Because signal0 and signal1 are transmitted at the same time into the air, they add to produce the raw signal:

$$(1, -1, -1, 1, 1, -1, 1, -1) + (-1, -1, -1, -1, 1, 1, 1, 1) = (0, -2, -2, 0, 2, 0, 2, 0).$$

This raw signal is called an interference pattern. The receiver then extracts an intelligible signal for any known sender by combining the sender's code with the interference pattern. The following table explains how this works and shows that the signals do not interfere with one another:

Step	Decode sender0	Decode sender1
0	code0 = $(1, -1)$, signal = $(0, -2, -2, 0, 2, 0, 2, 0)$	code1 = $(1, 1)$, signal = $(0, -2, -2, 0, 2, 0, 2, 0)$
1	decode0 = pattern.vector0	decode1 = pattern.vector1
2	decode0 = $((0, -2), (-2, 0), (2, 0), (2, 0)) \cdot (1, -1)$	decode1 = $((0, -2), (-2, 0), (2, 0), (2, 0)) \cdot (1, 1)$
3	decode0 = $((0 + 2), (-2 + 0), (2 + 0), (2 + 0))$	decode1 = $((0 - 2), (-2 + 0), (2 + 0), (2 + 0))$
4	data0 = $(2, -2, 2, 2)$, meaning $(1, 0, 1, 1)$	data1 = $(-2, -2, 2, 2)$, meaning $(0, 0, 1, 1)$

Further, after decoding, all values greater than 0 are interpreted as 1, while all values less than zero are interpreted as 0. For example, after decoding, data0 is $(2, -2, 2, 2)$, but the receiver interprets this as $(1, 0, 1, 1)$. Values of exactly 0 means that the sender did not transmit any data, as in the following example:

Assume signal0 = $(1, -1, -1, 1, 1, -1, 1, -1)$ is transmitted alone. The following table shows the decode at the receiver:

Step	Decode sender0	Decode sender1
0	code0 = $(1, -1)$, signal = $(1, -1, -1, 1, 1, -1, 1, -1)$	code1 = $(1, 1)$, signal = $(1, -1, -1, 1, 1, -1, 1, -1)$
1	decode0 = pattern.vector0	decode1 = pattern.vector1
2	decode0 = $((1, -1), (-1, 1), (1, -1), (1, -1)) \cdot (1, -1)$	decode1 = $((1, -1), (-1, 1), (1, -1), (1, -1)) \cdot (1, 1)$
3	decode0 = $((1 + 1), (-1 - 1), (1 + 1), (1 + 1))$	decode1 = $((1 - 1), (-1 + 1), (1 - 1), (1 - 1))$
4	data0 = $(2, -2, 2, 2)$, meaning $(1, 0, 1, 1)$	data1 = $(0, 0, 0, 0)$, meaning no data

When the receiver attempts to decode the signal using sender1's code, the data is all zeros, therefore the cross-correlation is equal to zero and it is clear that sender1 did not transmit any data.

Asynchronous CDMA

When mobile-to-base links cannot be precisely coordinated, particularly due to the mobility of the handsets, a different approach is required. Since it is not mathematically possible to create signature sequences that are both orthogonal for arbitrarily random starting points and which make full use of the code space, unique "pseudo-random" or "pseudo-noise" (PN) sequences are used in *asynchronous* CDMA systems. A PN code is a binary sequence that appears random but can be reproduced in a deterministic manner by intended receivers. These PN codes are used to encode and decode a user's signal in asynchronous CDMA in the same manner as the orthogonal codes in synchronous CDMA (shown in the example above). These PN sequences are statistically uncorrelated, and the sum of a large number of PN sequences results in *multiple access interference* (MAI) that is approximated by a Gaussian noise process (following the central limit theorem in statistics). Gold codes are an example of a PN suitable for this purpose, as there is low correlation between the codes. If all of the users are received with the same power level, then the variance (e.g., the noise power) of the MAI increases in direct proportion to the number of users. In other words, unlike synchronous CDMA, the signals of other users will appear as noise to the signal of interest and interfere slightly with the desired signal in proportion to number of users.

All forms of CDMA use spread-spectrum process gain to allow receivers to partially discriminate against unwanted signals. Signals encoded with the specified PN sequence (code) are received, while signals with different codes (or the same code but a different timing offset) appear as wideband noise reduced by the process gain.

Since each user generates MAI, controlling the signal strength is an important issue with CDMA transmitters. A CDM (synchronous CDMA), TDMA, or FDMA receiver can in theory completely reject arbitrarily strong signals using different codes, time slots or frequency channels due to the orthogonality of these systems. This is not true for asynchronous CDMA; rejection of unwanted signals is only partial. If any or all of the unwanted signals are much stronger than the desired signal, they will overwhelm it. This leads to a general requirement in any asynchronous CDMA system to approximately match the various signal power levels as seen at the receiver. In CDMA cellular, the base station uses a fast closed-loop power-control scheme to tightly control each mobile's transmit power.

Advantages of Asynchronous CDMA over other Techniques

Efficient Practical Utilization of the Fixed Frequency Spectrum

In theory CDMA, TDMA and FDMA have exactly the same spectral efficiency, but, in practice, each has its own challenges – power control in the case of CDMA, timing in the case of TDMA, and frequency generation/filtering in the case of FDMA.

TDMA systems must carefully synchronize the transmission times of all the users to ensure that they are received in the correct time slot and do not cause interference. Since this cannot be perfectly controlled in a mobile environment, each time slot must have a guard time, which reduces the probability that users will interfere, but decreases the spectral efficiency.

Similarly, FDMA systems must use a guard band between adjacent channels, due to the unpredictable Doppler shift of the signal spectrum because of user mobility. The guard bands will reduce the probability that adjacent channels will interfere, but decrease the utilization of the spectrum.

Flexible Allocation of Resources

Asynchronous CDMA offers a key advantage in the flexible allocation of resources i.e. allocation of PN codes to active users. In the case of CDM (synchronous CDMA), TDMA, and FDMA the number of simultaneous orthogonal codes, time slots, and frequency slots respectively are fixed, hence the capacity in terms of the number of simultaneous users is limited. There are a fixed number of orthogonal codes, time slots or frequency bands that can be allocated for CDM, TDMA, and FDMA systems, which remain underutilized due to the bursty nature of telephony and packetized data transmissions. There is no strict limit to the number of users that can be supported in an asynchronous CDMA system, only a practical limit governed by the desired bit error probability since the SIR (signal-to-interference ratio) varies inversely with the number of users. In a bursty traffic environment like mobile telephony, the advantage afforded by asynchronous CDMA is that the performance (bit error rate) is allowed to fluctuate randomly, with an average value determined by the number of users times the percentage of utilization. Suppose there are $2N$ users that only talk half of the time, then $2N$ users can be accommodated with the same *average* bit error probability as N users that talk all of the time. The key difference here is that the bit error probability for N users talking all of the time is constant, whereas it is a *random* quantity (with the same mean) for $2N$ users talking half of the time.

In other words, asynchronous CDMA is ideally suited to a mobile network where large numbers of transmitters each generate a relatively small amount of traffic at irregular intervals. CDM (synchronous CDMA), TDMA, and FDMA systems cannot recover the underutilized resources inherent to bursty traffic due to the fixed number of orthogonal codes, time slots or frequency channels that can be assigned to individual transmitters. For instance, if there are N time slots in a TDMA system and $2N$ users that talk half of the time, then half of the time there will be more than N users needing to use more than N time slots. Furthermore, it would require significant overhead to continually allocate and deallocate the orthogonal-code, time-slot or frequency-channel resources. By comparison, asynchronous CDMA transmitters simply send when they have something to say and go off the air when they don't, keeping the same PN signature sequence as long as they are connected to the system.

Spread-spectrum Characteristics of CDMA

Most modulation schemes try to minimize the bandwidth of this signal since bandwidth is a limited resource. However, spread-spectrum techniques use a transmission bandwidth that is several orders of magnitude greater than the minimum required signal bandwidth. One of the initial reasons for doing this was military applications including guidance and communication systems. These systems were designed using spread spectrum because of its security and resistance to jamming. Asynchronous CDMA has some level of privacy built in because the signal is spread using a pseudo-random code; this code makes the spread-spectrum signals appear random or have noise-like properties. A receiver cannot demodulate this transmission without knowledge of the pseudo-random sequence used to encode the data. CDMA is also resistant to jamming. A jamming signal only has a finite amount of power available to jam the signal. The jammer can either spread its energy over the entire bandwidth of the signal or jam only part of the entire signal.

CDMA can also effectively reject narrow-band interference. Since narrow-band interference affects only a small portion of the spread-spectrum signal, it can easily be removed through notch filtering without much loss of information. Convolution encoding and interleaving can be used to assist in recovering this lost data. CDMA signals are also resistant to multipath fading. Since the spread-spectrum signal occupies a large bandwidth, only a small portion of this will undergo fading due to multipath at any given time. Like the narrow-band interference, this will result in only a small loss of data and can be overcome.

Another reason CDMA is resistant to multipath interference is because the delayed versions of the transmitted pseudo-random codes will have poor correlation with the original pseudo-random code, and will thus appear as another user, which is ignored at the receiver. In other words, as long as the multipath channel induces at least one chip of delay, the multipath signals will arrive at the receiver such that they are shifted in time by at least one chip from the intended signal. The correlation properties of the pseudo-random codes are such that this slight delay causes the multipath to appear uncorrelated with the intended signal, and it is thus ignored.

Some CDMA devices use a rake receiver, which exploits multipath delay components to improve the performance of the system. A rake receiver combines the information from several correlators, each one tuned to a different path delay, producing a stronger version of the signal than a simple receiver with a single correlation tuned to the path delay of the strongest signal.

Frequency reuse is the ability to reuse the same radio channel frequency at other cell sites within a cellular system. In the FDMA and TDMA systems, frequency planning is an important consideration. The frequencies used in different cells must be planned carefully to ensure signals from different cells do not interfere with each other. In a CDMA system, the same frequency can be used in every cell, because channelization is done using the pseudo-random codes. Reusing the same frequency in every cell eliminates the need for frequency planning in a CDMA system; however, planning of the different pseudo-random sequences must be done to ensure that the received signal from one cell does not correlate with the signal from a nearby cell.

Since adjacent cells use the same frequencies, CDMA systems have the ability to perform soft hand-offs. Soft hand-offs allow the mobile telephone to communicate simultaneously with two or more cells. The best signal quality is selected until the hand-off is complete. This is different from hard hand-offs utilized in other cellular systems. In a hard-hand-off situation, as the mobile telephone approaches a hand-off, signal strength may vary abruptly. In contrast, CDMA systems use the soft hand-off, which is undetectable and provides a more reliable and higher-quality signal.

Collaborative CDMA

In a recent study, a novel collaborative multi-user transmission and detection scheme called collaborative CDMA has been investigated for the uplink that exploits the differences between users' fading channel signatures to increase the user capacity well beyond the spreading length in the MAI-limited environment. The authors show that it is possible to achieve this increase at a low complexity and high bit error rate performance in flat fading channels, which is a major research challenge for overloaded CDMA systems. In this approach, instead of using one sequence per user as in conventional CDMA, the authors group a small number of users to share the same spreading sequence and enable group spreading and despreading operations. The new collaborative multi-user receiver consists of two stages: group multi-user detection (MUD) stage to suppress the MAI between the groups and a low-complexity maximum-likelihood detection stage to recover jointly the co-spread users' data using minimal Euclidean-distance measure and users' channel-gain coefficients.

SPACE DIVISION MULTIPLE ACCESS

Space-division multiple access (SDMA) is a channel access method based on creating parallel spatial pipes next to higher capacity pipes through spatial multiplexing and/or diversity, by which it is able to offer superior performance in radio multiple access communication systems. In traditional mobile cellular network systems, the base station has no information on the position of the mobile units within the cell and radiates the signal in all directions within the cell in order to provide radio coverage. This method results in wasting power on transmissions when there are no mobile units to reach, in addition to causing interference for adjacent cells using the same frequency, so called co-channel cells. Likewise, in reception, the antenna receives signals coming from all directions including noise and interference signals. By using smart antenna technology and differing spatial locations of mobile units within the cell, space-division multiple access techniques offer attractive performance enhancements. The radiation pattern of the base station, both in transmission and reception, is adapted to each user to obtain highest gain in the direction of that user. This is often done using phased array techniques.

Use in Cellular Networks

GSM

In GSM cellular networks, the base station is aware of the distance (but not direction) of a mobile phone by use of a technique called "timing advance" (TA). The base transceiver station (BTS) can determine how far the mobile station (MS) is by interpreting the reported TA. This information, along with other parameters, can then be used to power down the BTS or MS, if a power control feature is implemented in the network. The power control in either BTS or MS is implemented in most modern networks, especially on the MS, as this ensures a better battery life for the MS. This is also why having a BTS close to the user results in less exposure to electromagnetic radiation.

This is why one may be safer to have a BTS close to them as their MS will be powered down as much as possible. For example, there is more power being transmitted from the MS than what one would receive from the BTS even if they were 6 meters away from a BTS mast. However, this estimation might not consider all the Mobile stations that a particular BTS is supporting with EM radiation at any given time.

5G

In the same manner, 5th generation mobile networks will be focused in using the given position of the MS in relation to BTS in order to focus all MS Radio frequency power to the BTS direction and vice versa, thus enabling power savings for the Mobile Operator, reducing MS SAR index, reducing the EM field around base stations since beam forming will concentrate RF power when it will be used rather than spread uniformly around the BTS, reducing health and safety concerns, enhancing spectral efficiency, and decreased MS battery consumption.

5

Modern Media for Communication

There are numerous tools which are used for communication such as social media, SMS text messaging, email marketing, direct email, blogging, voice calling, video chat, video marketing, live web chat and virtual reality. This chapter has been carefully written to provide an easy understanding of these methods of modern communication.

Social media has been around since the early days of the internet and it still dominates most of our lives. The long list of social networks continues to grow and each one is continuing to drive millions (and even billions) of users to their sites everyday which is why it is one of the most popular forms of communication.

Social media may be all about brand awareness, but it is also a great channel for communication as it enables you to post open messages for everyone to see, as well as engage with users through comments.

Whatever you use it for, it's a great first step for communication. Your messages aren't limited to your followers either, through the power of hashtags, shares, likes, retweets, hearts and other reaction your posts have no limit to the audience it can reach. We've all witnessed the power of posts going viral.

Social Media: Direct Message

Social media doesn't necessarily need to be completely public. Almost every social media channel offers a direct messaging option, some of those messaging services even have their own messaging app such as Facebook Messenger.

Private messaging through social networks has the same intimacy as email but tends to be less formal.

Businesses should only contact someone through Direct Message if the customer has reached out to them through that channel. However, certain direct messaging techniques are becoming a little more acceptable on some social networks such as Facebook. Facebook Messenger Bots are now seen as the norm and can help you to connect with your audience through Facebooks messaging service.

Instant Message (IM)

While some forms on Instant Message falls under social media like Facebook Messenger, there are a wide range of Instant Messaging platforms that aren't connected to social networks such as Google Hangouts and WhatsApp.

IM is a great tool for quick informal chats or group chats.

SMS Text Messaging

We are all aware of the uses of text messaging. They are short, generally informal and are a good way to communicate small bits of information that can be received and replied to at the recipient's own leisure.

These days we are almost always within arm's reach of a mobile device, so it isn't a surprise that more people are using their smartphones more than computers to search, research and communicate than ever before, making it the perfect platform to engage with your audience.

Email Marketing

The first mass email was sent back in 1978 and even back then it was highly successful. However, today it is one of the most underestimated marketing platforms around, despite being one of the most beneficial in terms of return on investment.

Email marketing can be used for many different purposes, including to push products and services, spread news, raise brand awareness or to deliver a message to the masses.

Over the years, most businesses would have accumulated a great deal of email address and in many cases, they would go to waste. With email marketing those email addresses will enable you to reconnect with old clients as well as connect with new potential clients.

Direct Email

Email is similar to direct messaging through social networks, but it is generally more formal. It is the most popular way of communicating between businesses with over 200 billion emails being sent every day.

Blogging

A blog is a conversational styled website that enables you to publish messages, news, knowledge or any other kind of information on the world wide web for everyone to see.

Most blogs include a comments section in which you can engage with those likeminded people that are interested in your blog post. This is why it is a great platform for communication.

Voice Calling

Voice calling is even more personalised than the channels previously mentioned. The telephone or mobile phone instantly allows both parties to hear the tones and emotions of the other caller and is one of the most commonly used communication tools.

Video Chat

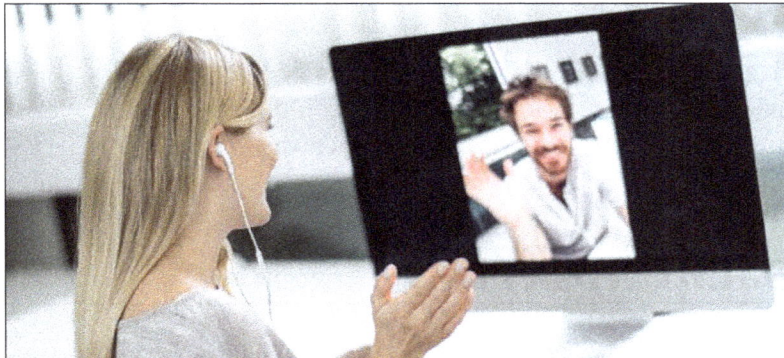

Video chat enables both parties to see each other, allowing you to be able to read body language and facial expressions. This form of communication isn't as popular as the voice calling but it does have its advantages.

With several video-calling apps available for free such as Apple's FaceTime, Facebook messenger, Skype and WhatsApp, video calls are definitely worth considering.

Video Marketing

Video has really taken off over the last few years with the help of social channels like YouTube, Facebook, Snapchat and Instagram. It also helps that it is easier than ever to record videos with smartphones and cameras always handily available.

It's important that you get your message across in a variety of formats and video is one of the most popular ways in which you can do so as it significantly boosts engagement.

Live Web Chat

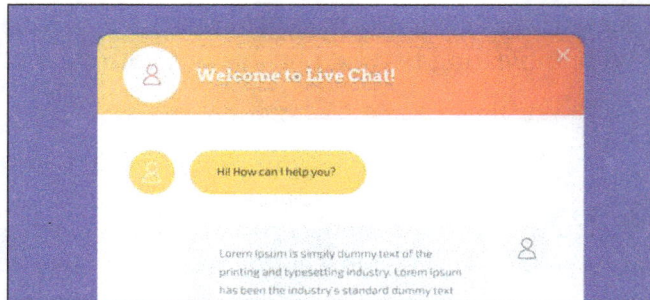

You are likely to have seen a lot of these on websites already and they can be very helpful if you are able to manage them correctly. Live web chats allow people on your website to ask you questions directly in real time without leaving your website.

You should only consider using live web chat if you can guarantee someone will always be actively using the widget. If it takes more than a few seconds to respond to someone via web chat it will reflect poorly on your business so it's best to avoid these widgets if you can't maintain the demand.

The video above shows us the potential virtual reality can have for communication. Facebook has been developing social VR for several years now and although it isn't a common way of communication yet, it has a lot of potential.

As virtual reality advances and prices for the headsets fall, VR popularity will inevitably increase. With more and more people buying VR headsets this could be the next modern form of communication to hit the masses and it may only be a few years away.

MOBILE COMMUNICATION

Mobile Communication is the use of technology that allows us to communicate with others in different locations without the use of any physical connection (wires or cables). Mobile communication makes our life easier, and it saves time and effort.

A mobile phone (also called mobile cellular network, cell phone or hand phone) is an example of mobile communication (wireless communication). It is an electric device used for full duplex two way radio telecommunication over a cellular network of base stations known as cell site.

Features of Mobile Communication

The following are the features of mobile communication:

- High capacity load balancing: Each wired or wireless infrastructure must incorporate high capacity load balancing.

High capacity load balancing means, when one access point is overloaded, the system will actively shift users from one access point to another depending on the capacity which is available.

- Scalability: The growth in popularity of new wireless devices continuously increasing day by day. The wireless networks have the ability to start small if necessary, but expand in terms of coverage and capacity as needed - without having to overhaul or build an entirely new network.

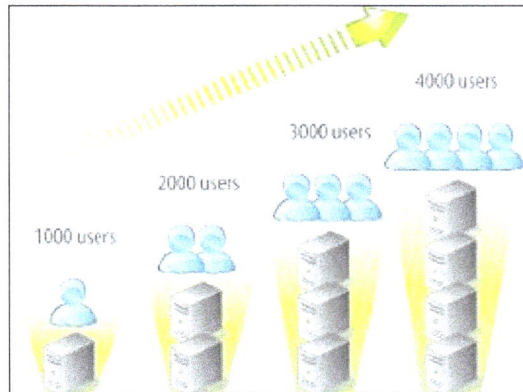

- Network management system: Now a day, wireless networks are much more complex and may consist of hundreds or even thousands of access points, firewalls, switches, managed power and various other components.

The wireless networks have a smarter way of managing the entire network from a centralized point.

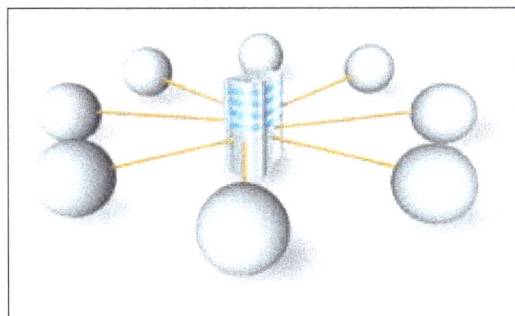

- Role based access control (RBAC) allows you to assign roles based on what, who, where, when and how a user or device is trying to access your network.

Once the end user or role of the devices is defined, access control policies or rules can be enforced.

- Indoor as well as outdoor coverage options: It is important that your wireless system has the capability of adding indoor coverage as well as outdoor coverage.

- Network access control: Network access control can also be called as mobile device registration. It is essential to have a secure registration.

- Network access control (NAC) controls the role of the user and enforces policies. NAC can allow your users to register themselves to the network. It is a helpful feature that enhances the user experience.

- Mobile device management: Suppose, many mobile devices are accessing your wireless network; now think about the thousands of applications are running on those mobile devices.

How do you plan on managing all of these devices and their applications, especially as devices come and go from your business?

Mobile device management can provide control of how you will manage access to programs and applications. Even you can remotely wipe the device if it is lost or stolen.

- Roaming: You don't need to worry about dropped connections, slower speeds or any disruption in service as you move throughout your office or even from building to building wireless needs to be mobile first.

Roaming allows your end-users to successfully move from one access point to another without ever noticing a dip in a performance.

For example, allowing a student to check their mail as they walk from one class to the next.

- Redundancy: The level or amount of redundancy your wireless system requires depends on your specific environment and needs.

- For example: A hospital environment will need a higher level of redundancy than a coffee shop. However, at the end of the day, they both need to have a backup plan in place.

- Proper Security means using the right firewall: The backbone of the system is your network firewall. With the right firewall in place you will be able to:

- See and control both your applications and end users.

- Create the right balance between security and performance.

- Reduce the complexity with:

  ◦ Antivirus protection.

  ◦ Deep Packet Inspection (DPI).

  ◦ Application filtering.

- Protect your network and end users against known and unknown threads including:

  ◦ Zero- day.

  ◦ Encrypted malware.

  ◦ Ransomware.

  ◦ Malicious botnets.

- Switching: Basically, a network switch is the traffic cop of your wireless network which making sure that everyone and every device gets to where they need to go.

Switching is an essential part of every fast, secure wireless network for several reasons:

  ◦ It helps the traffic on your network flow more efficiently.

  ◦ It minimizes unnecessary traffic.

  ◦ It creates a better user experience by ensuring your traffic is going to the right places.

Advantages of Mobile Communication

There are following advantages of mobile communication:

- Flexibility: Wireless communication enables the people to communicate with each other regardless of location. There is no need to be in an office or some telephone booth in order to pass and receive messages.

- Cost effectiveness: In wireless communication, there is no need of any physical infrastructure (Wires or cables) or maintenance practice. Hence, the cost is reduced.

- Speed: Improvements can also be seen in speed. The network connectivity or the accessibility was much improved in accuracy and speed.

- Accessibility: With the help of wireless technology easy accessibility to the remote areas is possible. For example, in rural areas, online education is now possible. Educators or students no longer need to travel to far-flung areas to teach their lessons.

- Constant connectivity: Constant connectivity ensures that people can respond to emergencies relatively quickly. For example, a wireless device like mobile can ensure you a constant connectivity though you move from place to place or while you travel, whereas a wired landline can't.

Cellular Concept

The advent of cellular operation brought frequency reuse capabilities. Advances in wireless access, digital signal processing, integrated circuits, increased battery life, etc led to exponential growth of personal communication services.

The cellular system works as follows: An available frequency spectrum is divided into discrete channels, which are assigned in groups to geographic cells covering a service area. The discrete channels are capable of being reused in different cells with diameters ranging from 2 to 50 km. The service area is allotted a radio frequency (RF) transmitter, whereas adjacent cells operate on different frequencies to avoid interference.

Cellular telephones began as a simple two-way analogue communication system using frequency modulation for voice and frequency-shift keying for transporting control and signalling information. Other cellular systems are a digital cellular system, cordless telephony, satellite mobile and paging. Analogue cellular systems fall in the first-generation(1G) category and digital cellular low-power wireless fall in the second-generation (2G) category.

Analogue Cellular Phone

In 1970, Bell Labs in New Jersey proposed a cellular telephone concept as advanced mobile telephony system (AMPS). AMPS is a standard cellular telephone service placed into operation on October 13, 1983, by Illinois Bell. It uses narrow-band FM with a usable audio frequency band of 300-3 kHz and maximum frequency deviation of ±12 kHz for 100 per cent modulation. According to Carson's rule, this corresponds to 30 kHz.

AMPS uses frequency-division multiple access (FDMA), where transmissions are separated in the frequency domain. Subscribers are assigned a pair of voice channels (forward and reverse) for the duration of their call. Analogue cellular channels carry both voices using FM and digital signalling information using binary FSK.

Digital Cellular System

It provides improvements in both capacity and performance. FDMA uses a frequency canalisation approach to spectrum management, while time-division multiple access (TDMA) utilises a time-division approach. The entire available cellular RF spectrum is sub-divided into narrow-band radio channels to be used as a one-way communication link between cellular mobile units and base stations.

Multiple Access Technologies for Cellular Systems

Generally, a fixed amount of frequency spectrum is allocated to a cellular system. Multiple access techniques are deployed so that the users can share the available spectrum in an efficient manner.

For wireless communication, multiplexing can be carried out in three dimensions: Time (TDMA), frequency (FDMA and its variation OFDMA) and code (CDMA).

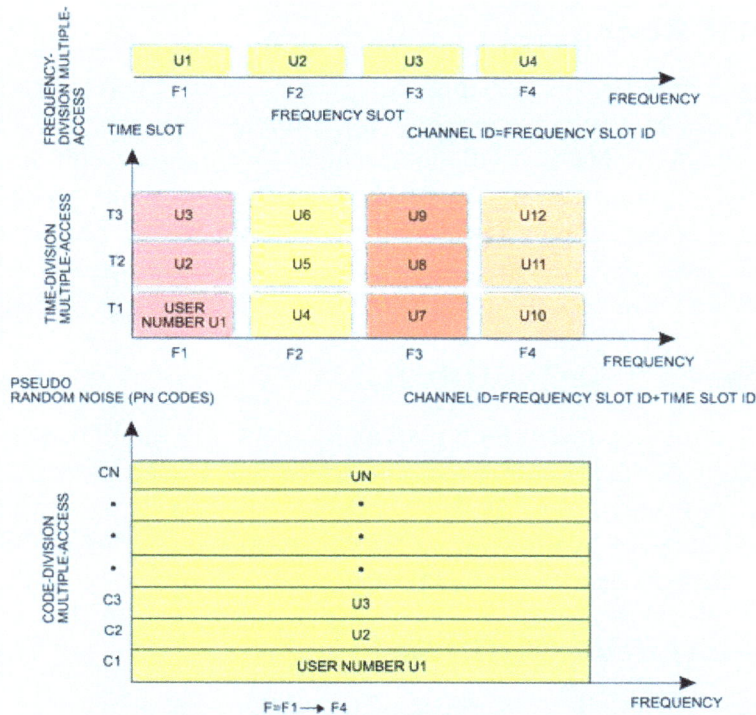

FDMA, TDMA, and CDMA multiple-access techniques.

In TDMA the available spectrum is partitioned into narrow frequency bands or frequency channels, which, in turn, are divided into a number of time slots. In case of North American digital cellular standard IS-136, each frequency channel (30 kHz) is divided into three-time slots, whereas in European digital cellular system GSM each frequency channel (200 kHz) is divided into eight time slots. Guard bands are needed both between frequency channels and time slots.

In FDMA, users share the available spectrum in a frequency band called traffic channel. Different users are assigned different channels on demand basis. The user's signal power is concentrated in a relatively narrow frequency band. All the analogue cellular systems used FDMA system.

OFDM is a multi-cellular transmission technique where a data stream is carried with many lower-rate subcarrier tones. It has been adopted in mobile communications to combat hostile frequency-selective fading and has been incorporated into wireless network standards. OFDM is a multi-cellular transmission technique where a data stream is carried with many lower-rate sub-carrier tones. It has been adopted in mobile communications to combat hostile frequency-selective fading and has been incorporated into wireless network standards.

OFDM combines the advantages of coherent detection and OFDM modulation and has many merits that are critical for future high-speed transmission systems. By using up/down conversion, electrical bandwidth requirement can be greatly reduced for the OFDM transceiver, which is extremely attractive for high-speed circuit design where electrical signal bandwidth dictates the cost. Lastly, signal processing in the OFDM transceiver can take advantage of an efficient algorithm of fast Fourier transform (FFT)/inverse FFT, which suggests that OFDM has superior scalability over channel dispersion and data rate.

Digital Modulation Keying

Communication systems often involve the modulation of a carrier, which results in a bandpass wave-form. A digital signal can be used to modulate the amplitude, frequency or phase of a sinusoidal carrier producing three different forms of digital modulation: amplitude-shift keying (ASK), frequen-cy-shift keying (FSK) and phase-shift keying (PSK). In addition to these basic techniques, there are some modulation schemes that employ a combination of amplitude and phase modulation. It may be noted that unlike ASK signal, PSK transmission is polar. At the same time, ASK is a linear modulation scheme, whereas PSK is a non-linear modulation scheme. PSK has a superior performance over ASK.

Quadrature Phase-shift Keying (QPSK)

Digital modulation techniques mentioned above are spectrally inefficient in the sense that the available channel bandwidth is not fully used. Spectral efficiency can be improved by using QPSK. It is a system for two message sources. In this system modulation carriers in phase quadrature are combined to form the output waveform. In QPSK the amplitude of the modulator waveform and modulator gains are made as nearly equal as possible.

Differential Phase-shift keying (DPSK)

DPSK is a modification of PS that avoids the need to provide synchronous carrier required for de-tection of PSK signals. It is an ingenious technique whereby the carrier reference is derived from the received waveform in the preceding bit interval by use of a 1-bit delay. In essence, the received waveform delayed by 1-bit duration serves as its own reference.

Data Transmission using Packet Switching

This is done by supplying various addressed packets, which are interconnected to have the conver-sation. New dedicated paths are created for sending the data. From the multiple paths to the des-tination, any path can be used to send data. Cellular digital packet data was designed for optimal operation with an analogue cellular system, especially AMPS.

Short Message Service

Short message service is the most common packet service that is supported on digital cellular net-works like GSM, IS-136, EDGE and PDC (packet data service). It is a store-and-forward/packet mode service that provides inter-working with the various applications and services within a fixed network. For message transfer between relevant network entities, control and signalling channels (instead of normal traffic channels) are generally used for data transmission.

General Packet Radio Service (GPRS)

GPRS essentially represents add-on capabilities to the basic voice-optimised cellular network that nevertheless maintain the essential characteristics of radio-access technology. You can use these using GPRS or GSM modules.

Enhanced Data Rates for GSM Evolution (EDGE)

In order to enhance the data handling capabilities of 2G service, the radio-access portion had to be modified. This modification was evolved in Europe in the form of EDGE. EDGE also supports a link adaptation mechanism that selects the best combination of modulation and encoding schemes based on the time varying link quality.

EDGE concept applies to both circuit-mode and packet-mode data and is sufficiently generic for application to other digital cellular systems. It works in the 200kHz bandwidth with one or more high-level modulation schemes and a range of efficient coding methods. Modulation schemes are offset QPSK and offset 16 QAM.

Spread Spectrum

It is a special communication technique that purposefully uses much more RF bandwidth than necessary to transmit a signal. This helps in improving the signal-to-noise (S/N) ratio. The main advantages of this technique are secure communication and resistance to intentional jamming. There are 75 channels in the 2400-2483.3MHz band.

Methods of Performing Spread Spectrum

Frequency Hopping

This technique spreads the narrow-band signal as a function of time. The transmitted frequency is changed to a different pre-assigned channel several times per second (hopped). The order in which the pre-assigned channels are selected is 'pseudo random.' In other words, the channel order is seemingly random but actually repeats itself at a definedinterval. The specificorder in which frequencies are occupied is a function of code sequence and the rate of hopping from one frequency to another is a function of information rate.

Direct Sequence

This technique spreads a signal by expanding the signal over a broadband portion of the radio band. It uses a locally generated pseudo noise (PN) code to encode digital data to be transmitted. The most practical all-digit version is direct sequence. Binary phase-shift keying is the simplest and most often used modulation technique.

One of the most important features of spread-spectrum signals is that these contain a large number of very different signaling formats, used for communicating data symbols. It means that the receiver which detects one of these formats cannot detect any other format within a single message. The number of formats used in a spread-spectrum system is called multiplicity factor of the communication link and amounts to thousands.

CDMA

CDMA is a form of direct-sequence spread-spectrum technology that allows many users to occupy the same time and frequency allocations in a given band/space. CDMA assigns each user a unique spreading code to spread the baseband data before transmission, in order to help differentiate signals from various users in the same spectrum. It is the platform on which 2G and advanced 3G services are built.

After speech, the codec converts voice into digital signal. CDMA spreads the voice stream over the full 1.25MHz bandwidth of the CDMA channel, coding each stream separately. The receiver uses a correlator to despread the wanted signal, which is passed through a bandpass filter.Unwanted signals are not despread and not passed through the filter.

The rate of the spreading signal is known as the 'chip rate' as each bit in the spreading signal is known as 'chip.' All 2G networks support only single-user data rates of the order of 10 kbps, which is too slow for rapid e-mail and Internet browsing.

CDMA provides more than ten times the capacity of the analogue AMPS and fivetimes the calling capacity of GSM and TDMA systems. It requires fewer cell sites than GSM and TDMA.

Personal Communication System

Personal communication system (PCS) is a new class of cellular telephone system such as AMPS. PCS systems are a combination of cellular telephone network and intelligent network, which is the entity of super-simple transfer (SST) inter-office protocol tha distinguishes physical components of the switching network such as signal service point, signal control point and signal transfer point from the services provided by SST network.

In essence, PCS is the North American implementation of European GSM standard. GSM utilised its own TDMA access methods and provided expanded capacity and unique services such as caller ID, call forwarding and short messaging. A critical feature was seamless roaming, which allowed subscribers to move across provider boundaries. The effort was directed towards second-generation cellular systems.

In 1990, a second frequency band was specified. This band included twodomains—1710-1785 MHz and 1805-1880 MHz, i.e., twice 75 MHz; three times as much as the primary 900MHz band.

Digital Enhanced Cordless Telecommunication (DECT)

DECT is a type of PCS system. DECT standard was developed by European Telecommunication Standards Institute (ETSI) for wireless PABX data LAN applications that represent closed environments requiring minimal open cordless access, since it was essential that products from different vendors not only coexist but interwork with each other.

DECT system has a TDMA/TDD frame structure with 24 slots that are equally allocated for downlink and uplink operation. DECT specifiesboth simplex (half-slots) and duplex (full slot) operation. Higher data rates are achieved by utilising multilevel modulation. The basic modulation scheme is a two-level Gaussian filled frequency-shif keying (GFSK), which is supplemented with 8-level modulation scheme leading to as high as 2.88 Mbps per carrier.

GSM

Global system for mobile communications (GSM) was developed by the Groupe Special Mobile, which was an initiative of the Conference of European Post and Telecommunications (CEPT) administrations. GSM was firs devised as a cellular system in a specific 900MHz band called the primary band. This primary band includes two sub-bands of 25 MHz each, 890-915 MHz and 935-960 Mhz.

GSM systems like Iridium, Globalstar and ICO use constellations of low-earth orbit (LEO) or medium-earth orbit (MEO) satellites and operate as overlay networks for existing cellular and PCS networks. Using dual-mode, these extend the coverage to any and all locations on the earth's surface.

GPS—a reliable navigational aid anywhere on the earth.

International Mobile Telecommunication-2000 (IMT-2000) is a standard developed by ITU for 3G. It ensures global mobility in terms of global seamless roaming and service delivery. An appreciation of the role of numbering and identities in mobility management, international roaming, call delivery, and billing and charging is important in understanding the operation of mobile and personal communication networks.

Personal communication satellite service (PCSS) uses LEO satellite repeaters incorporating QPSK modulation and both FDMA and TDMA.

The main advantages of GSM are international roaming (in harmony with ISDN principles assuring inter-working between ISDN and GSM) and features like privacy and encryption, frequency hopping, discontinuous transmission and short message service. Other facilities include call forwarding, barring, waiting, hold and teleconferencing.

The basic architecture comprises a network sub-system, base station sub-system, mobile stations, and system interworking and interfaces.

A subscriber identity module (SIM) is required to activate and operate a GSM terminal. The SIM may be contained within the mobile station or it may be a removable unit that can be inserted by the user in his mobile set.

New Developments Along the way

Before we proceed to evolution from 1G to 4G, let me touch upon the new developments that took place in 1G to 4G.

Global Positioning System (GPS)

GPS is a reliable navigational aid available anywhere on the earth, operating in all weather conditions 24 hours a day. It can be used by marine, airborne and land users. GPS technology was developed in 1983.

Segments of GPS

Space Segment

GPS consists of 24 NAVSTAR satellites along with three spare satellites orbiting at 20,200 km above the earth's surface in six circular orbital planes with a 12-hour orbital period each. These satellites operate at L1 band (1.575 GHz) continuously broadcasting navigational signals called coarse acquisition code. These codes can be received by anyone for decoding and findingnavigational parameters like longitude, latitude, velocity and time.

Control Segment

It consists of a master control station (MCS) and a number of smaller earth stations called monitoring stations located at different places in the world. Monitoring stations track satellites and pass on the measured data to the MCS. The MCS computes satellite parameters (called ephemeris) and sends them back to the satellite, which, in turn, broadcasts to all GPS receivers.

User Segment

The user segment consists of all moving and stationary objects with GPS receivers. A GPS receiver is a multi-channel satellite receiver that computes every second its own location and velocity.

Bluetooth

Compared to WLAN technologies, Bluetooth technology aims at so-called ad-hoc piconets, which are local-area networks with a very limited coverage and without the need for an infrastructure. The term 'piconet' is a collection of Bluetooth devices that are synchronised to the same hopping sequence. One device in the piconet can act as master and all other devices connected to the master act as slaves. The master determines the hopping pattern and the slaves have to synchronise to this pattern. The hopping pattern is determined by the device ID—a 48-bit worldwide unique identifier.The phase in the hopping pattern is determined by the master's clock. All active devices are assigned a 3-bit active member address.

All parked devices use an 8-bit parked member address. Devices in standby mode do not need an address. The goal for Bluetooth development was to use a single-chip, low-cost, radio-based wireless network technology for laptops, notebooks, headsets, etc.

Bluetooth operates in the 2.4GHz ISM band. However, MAC, physical layer and the offered services are completely different. Bluetooth transceivers use Gaussian FSK for modulation and are available in three power classes: Class 1 (max. power 100 mW), class 2 (max. power 2.5 mW) and class 3 (max. power 1 mW).

Journey from 1G to 4G

1G System

1G specifications were released in 1990 to be used in GSM. 1G systems are analogue systems such as AMPS that use FDM to divide the bandwidth into specificfrequencies that are assigned to individual calls.

2G System

These second-generation mobile systems are digital and use either TDMA or CDMA method. Digital cellular systems use digital modulation and have several advantages over analogue systems, including better utilisation of bandwidth, more privacy, and incorporation of error detection and correction.

2.5G System

It was introduced mainly to add latest bandwidth technology to the existing 2G generation. It supports higher-data-rate transmission for Web browsing and also supports a new browsing format language called wireless application protocol (WAP). The different upgrade paths include high-speed circuit-switched data (HSCSD), GPRS and EDGE.

HSCSD increases the available application data rate to 14.4 kbps as compared to 9.6 kbps of GSM. By using four consecutive time slots, HSCSD is able to provide a raw transmission rate of up to 57.6 kbps to individual users.

GPRS supports multi-user network sharing of individual radio channels and time slots. Thus GPRS supports many more users than HSCSD but in a bursty manner. When all the eight time slots of a GSM radio channel are dedicated to GPRS, an individual can achieve as much as 171.2 kbps. But this has not brought any new evolution.

EDGE introduces a new digital modulation format called 8-PSK (octal phase-shift keying). It allows nine different air interface formats, known as multiple modulation and coding schemes, with varying degree of error control and protection. These formats are automatically and rapidly selectable. Of course, the covering range is smaller in EDGE than in HSCSD or GRPS.

3G System

To overcome the short-comings of 2G and 2.5G, 3G has been developed. It uses a wideband wireless network that offers increased clarity in conversations. Countries throughout the world are currently determining new radio spectrum bands to accommodate 3G networks. ITU has established 2500-2690MHz, 1700-1855MHz and 806-960MHz bands. Here the target data rate is 2 Mbps. The data is sent through packet switching. Voice calls are interpreted through circuit switching.

3G W-CDMA (UMTS)

Universal Mobile Telecommunication System (UMTS) or W-CDMA assures backward compatibility with 2G and 2.5G TDMA technologies. W-CDMA, which is an air interface standard, has been designed for always-on packet-based wireless service, so that computers and entertainment devices may all share the same wireless network and connect to the Internet anytime, anywhere.

W-CDMA supports data rates of up to 2.048 Mbps if the user is stationary, thereby allowing high-quality data, multimedia, streaming audio, streaming video and broadcast type services to consumers. With W-CDMA, data rates from as low as 8 kbps to as high as 2 Mbps can be carried simultaneously on a single W-CDMA 5MHz radio channel, with each channel supporting between 100 and 350 simultaneous voice calls at once, depending on antenna sectoring, propagation conditions, user velocity and antenna polarisation.

Time slots in W-CDMA are not used for user separation but to support periodic functions. (This is in contrast to GSM where time slots are used to separate users). The bandwidth per W-CDMA channel is 4.4 to 5 MHz.

Since the global standard was difficult to evolve, three operating modes have been specified:A 3G device will be a personal, mobile, multimedia communication device (e.g., TV provider redirects a TV channel directly to the subscriber's phone where it can be watched). Second, it will support video conferencing, i.e., subscribers can see as well as talk to each other. Third, it will also support location-based services, where a service provider sends localised weather or trafficconditions to the phone or the phone allows the subscriber to findnearby businesses or friends.

3.5G

It supports a higher through-put and speed at packet data rates of 14.4 Mbps, supporting higher data needs of consumers.

4G system

It offers additional features such as IP telephony, ultrabroadband Internet access, gaming services and HDTV streamed multimedia. Flash-OFDM, the 802.16e mobile version of WiMax (also known as WiBro in South Korea), can support cellular peak data rates of approx. 100 Mbps for high-mobility communications such as mobile access and up to 1 Gbps for low-mobility communications such as nomadic/local wireless access, using scalable bandwidths of up to 40 MHz. The infrastructure for 4G is only packet-based (all-IP).

SATELLITE COMMUNICATION

A satellite is a body that moves around another body in a mathematically predictable path called an Orbit. A communication satellite is nothing but a microwave repeater station in space that is helpful in telecommunications, radio, and television along with internet applications.

A repeater is a circuit which increases the strength of the signal it receives and retransmits it. But here this repeater works as a transponder, which changes the frequency band of the transmitted signal, from the received one.

The frequency with which the signal is sent into the space is called Uplink frequency, while the frequency with which it is sent by the transponder is Downlink frequency.

The following figure illustrates this concept clearly.

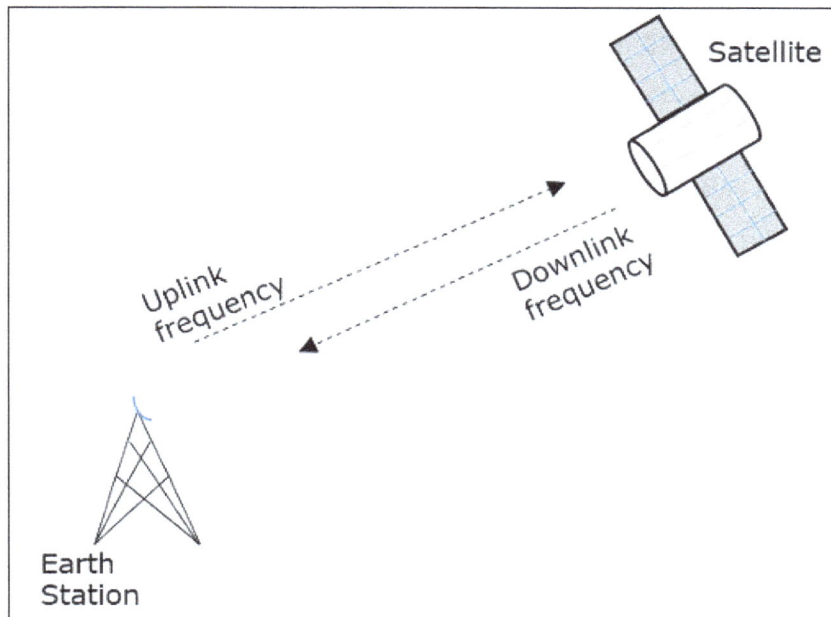

Now, let us have a look at the advantages, disadvantages and applications of satellite communications.

Advantages

There are many Advantages of satellite communications such as:

- Flexibility.
- Ease in installing new circuits.
- Distances are easily covered and cost doesn't matter.
- Broadcasting possibilities.
- Each and every corner of earth is covered.
- User can control the network.

Disadvantages

Satellite communication has the following drawbacks:

- The initial costs such as segment and launch costs are too high.

- Congestion of frequencies.

- Interference and propagation.

Applications

Satellite communication finds its applications in the following areas:

- In Radio broadcasting.

- In TV broadcasting such as DTH.

- In Internet applications such as providing Internet connection for data transfer, GPS applications, Internet surfing, etc.

- For voice communications.

- For research and development sector, in many areas.

- In military applications and navigations.

The orientation of the satellite in its orbit depends upon the three laws called as Kepler's laws.

Kepler's Laws

Johannes Kepler the astronomical scientist, gave 3 revolutionary laws, regarding the motion of satellites. The path followed by a satellite around its primary (the earth) is an ellipse. Ellipse has two foci - F1 and F2, the earth being one of them.

If the distance from the center of the object to a point on its elliptical path is considered, then the farthest point of an ellipse from the center is called as apogee and the shortest point of an ellipse from the center is called as perigee.

Kepler's 1st Law

Kepler's 1st law states that, "every planet revolves around the sun in an elliptical orbit, with sun as one of its foci." As such, a satellite moves in an elliptical path with earth as one of its foci.

The semi major axis of the ellipse is denoted as 'a' and semi minor axis is denoted as b. Therefore, the eccentricity e of this system can be written as:

$$e = \sqrt{\frac{a^2 - b^2}{a}}$$

- Eccentricity (e) – It is the parameter which defines the difference in the shape of the ellipse rather than that of a circle.

- Semi-major axis (a) – It is the longest diameter drawn joining the two foci along the center, which touches both the apogees (farthest points of an ellipse from the center).

- Semi-minor axis (b) – It is the shortest diameter drawn through the center which touches both the perigees (shortest points of an ellipse from the center).

These are well described in the following figure.

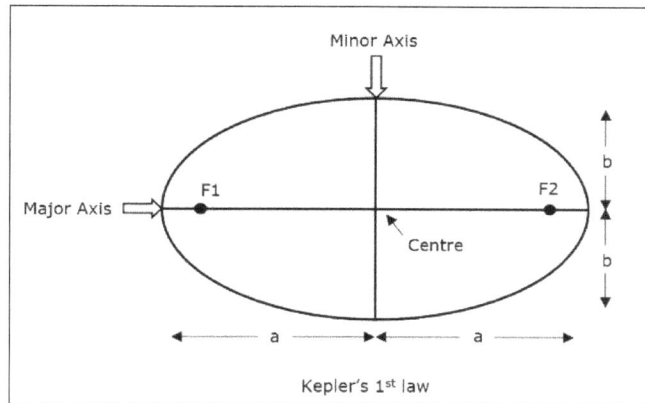

Kepler's 1st law

For an elliptical path, it is always desirable that the eccentricity should lie in between 0 and 1, i.e. 0 < e < 1 because if e becomes zero, the path will be no more in elliptical shape rather it will be converted into a circular path.

Kepler's 2nd Law

Kepler's 2nd law states that, "For equal intervals of time, the area covered by the satellite is equal with respect to the center of the earth."

It can be understood by taking a look at the following figure.

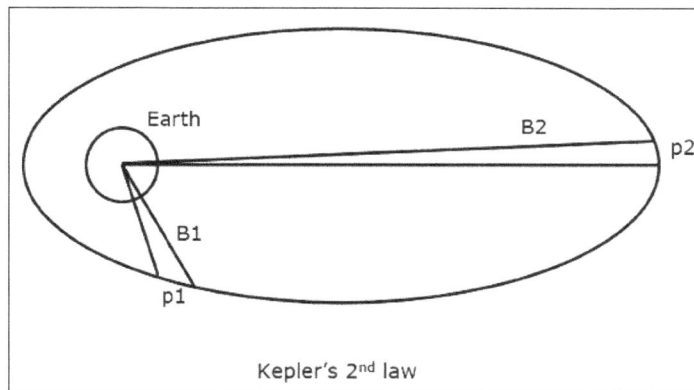

Kepler's 2nd law

Suppose that the satellite covers p1 and p2 distances, in the same time interval, then the areas B1 and B2 covered in both instances respectively, are equal.

Kepler's 3rd Law

Kepler's 3rd law states that, "The square of the periodic time of the orbit is proportional to the cube of the mean distance between the two bodies."

This can be written mathematically as,

$$T^2 \alpha \, a^3$$

which implies,

$$T^2 = \frac{4\pi^2}{GM}a^3$$

where $\frac{4\pi^2}{GM}$ is the proportionality constant (according to Newtonian Mechanics),

$$T^2 = \frac{4\pi^2}{\mu}a^3$$

where μ = the earth's geocentric gravitational constant, i.e. M = 3.986005 × 10^{14} m^3 / sec^2

$$1 = (\frac{2\pi}{T})^2 \frac{a3}{\mu}$$

$$1 = n^2 \frac{a^3}{\mu} \Rightarrow a^3 = \frac{\mu}{n^2}$$

where n = the mean motion of the satellite in radians per second.

The orbital functioning of satellites is calculated with the help of these Kepler's laws.

Along with these, there is an important thing which has to be noted. A satellite, when it revolves around the earth, undergoes a pulling force from the earth which is the gravitational force. Also, it experiences some pulling force from the sun and the moon. Hence, there are two forces acting on it. They are:

- Centripetal force – The force that tends to draw an object moving in a trajectory path, towards itself is called as centripetal force.

- Centrifugal force – The force that tends to push an object moving in a trajectory path, away from its position is called as centrifugal force.

So, a satellite has to balance these two forces to keep itself in its orbit.

Earth Orbits

A satellite when launched into space, needs to be placed in a certain orbit to provide a particular way for its revolution, so as to maintain accessibility and serve its purpose whether scientific, military, or commercial. Such orbits which are assigned to satellites, with respect to earth are called as Earth Orbits. The satellites in these orbits are Earth Orbit Satellites.

The important kinds of Earth Orbits are:

- Geo Synchronous Earth Orbit.

- Medium Earth Orbit.

- Low Earth Orbit.

Geosynchronous Earth Orbit Satellites

A Geo-synchronous Earth Orbit (GEO) satellite is one which is placed at an altitude of 22,300 miles above the Earth. This orbit is synchronized with a side real day (i.e., 23hours 56minutes). This orbit can have inclination and eccentricity. It may not be circular. This orbit can be tilted at the poles of the earth. But it appears stationary when observed from the Earth.

The same geo-synchronous orbit, if it is circular and in the plane of equator, it is called as geo-stationary orbit. These satellites are placed at 35,900kms (same as geosynchronous) above the Earth's Equator and they keep on rotating with respect to earth's direction (west to east). These satellites are considered stationary with respect to earth and hence the name implies.

Geo-Stationary Earth Orbit Satellites are used for weather forecasting, satellite TV, satellite radio and other types of global communications.

The following figure shows the difference between Geo-synchronous and Geo-stationary orbits. The axis of rotation indicates the movement of Earth.

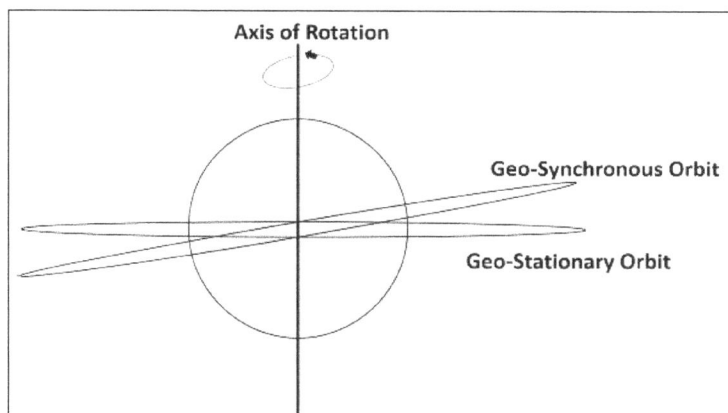

Every geo-stationary orbit is a geo-synchronous orbit. But every geo-synchronous orbit is not a Geo-stationary orbit.

Medium Earth Orbit Satellites

Medium Earth Orbit (MEO) satellite networks will orbit at distances of about 8000 miles from the earth's surface. Signals transmitted from a MEO satellite travel a shorter distance. This translates to improved signal strength at the receiving end. This shows that smaller, more lightweight receiving terminals can be used at the receiving end.

Since the signal is travelling a shorter distance to and from the satellite, there is less transmission delay. Transmission delay can be defined as the time it takes for a signal to travel up to a satellite and back down to a receiving station.

For real-time communications, the shorter the transmission delay, the better will be the communication system. As an example, if a GEO satellite requires 0.25 seconds for a round trip, then MEO satellite requires less than 0.1 seconds to complete the same trip. MEOs operates in the frequency range of 2 GHz and above.

Low Earth Orbit Satellites

The Low Earth Orbit (LEO) satellites are mainly classified into three categories namely, little LEOs, big LEOs, and Mega-LEOs. LEOs will orbit at a distance of 500 to 1000 miles above the earth's surface.

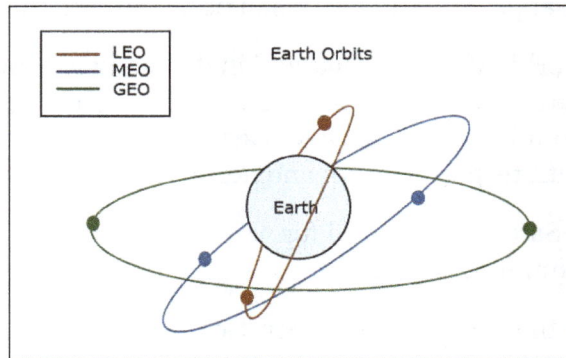

This relatively short distance reduces transmission delay to only 0.05 seconds. This further reduces the need for sensitive and bulky receiving equipment. Little LEOs will operate in the 800 MHz (0.8 GHz) range. Big LEOs will operate in the 2 GHz or above range, and Mega-LEOs operates in the 20-30 GHz range.

The higher frequencies associated with Mega-LEOs translates into more information carrying capacity and yields to the capability of real-time, low delay video transmission scheme.

The following figure depicts the paths of LEO, MEO, and GEO.

6

Diverse Aspects of Digital Communication

There are numerous aspects of digital communications such as bit rate, baud rate error detection and correction code, bandwidth, signal to noise ratio, modulation error ratio, Shannon-Hartley theorem, etc. This chapter closely examines these aspects of digital communications to provide an extensive understanding of the subject.

BIT RATE

Bit rate (sometimes written bitrate) is the number of bits of data that are conveyed per unit of time. The bit rate is quantified using the bits per second unit (symbol bit/s), often in conjunction with an SI prefix such as kilo (kbit/s), mega (Mbit/s), giga (Gbit/s) or tera (Tbit/s). One byte per second (1 B/s) corresponds to 8 bit/s. This task is done by the bit rate controller, an algorithm which dynamically adjusts encoder parameters to achieve a desired bit rate. It allocates a budget of bits to each picture or sub-parts of picture in a video sequence. It is not a part of the H.264 standard.

Ideal Behavior

We will as such focus on two simple examples:

- Low complexity scene (with little/no motion).

- High complexity scene (with medium/big motion).

Variable Bit Rate

Variable bit rate lets the video use the bandwidth it needs. The bandwidth can be really high or really low depending on the scene's complexity. VBR allows a high bitrate (and therefore requires more storage space and a network capable of handling it) to be allocated for more complex scenes of a video while using less bit rate for less complex scenes. The average of these rates is then calculated to get the average bitrate for the sequence. The advantages of VBR are that it produces a better visual quality on a video (has a better rate-distortion ratio) compared to Constant Bit Rate or Maximum Bit Rate for the same scene. The disadvantages are that the bit rate of the video may be really big in complex sequences. VBR may also pose problems during streaming when the instantaneous bitrate exceeds the available bandwidth of the network or the capacity of the recording/playing devices. Ideally, Variable Bit Rate doesn't require any bandwidth regulation and uses as much bandwidth as needed.

The quality is constant whatever happens on the video:

- Low complexity scene: The bit rate will be low, but as soon as something occurs then the bit rate will jump to a higher level.

- High complexity scene: The bit rate will be high and may overload the recording device, network or video player.

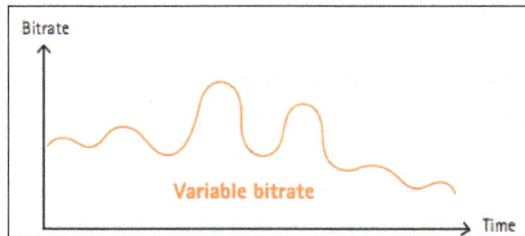

Constant Bit Rate

Constant bit rate aims at having a bit rate for which a encoder's output data should be consumed is constant, does not evolve over time and stays at the desired target bit rate whatever happens on the video. CBR is useful for streaming multimedia content on limited capacity channels. It was first designed for the video industry (VCD, DVD, Blu-ray) to ensure that the bandwidth for every device wouldn't be exceeded and to ensure a continuous playback.

This mode is less good for storage because it may contain padding data and waste storage space for no gain on the video quality. If the video is not complex enough to sustain the target bit rate, the encoder will complete the remaining bit rate gap by doing bit padding: filling the video with unnecessary information (information not part of the video).

To meet the bitrate target, the bit rate controller changes the video quality and as last option reduce the frame rate.

- Low complexity scene: The bit rate will be the one set as target, bit padding may occur as we may not have enough data to achieve the target. Video will have full frame rate.

- High complexity scene: The bit rate will still be one set as target but the visual quality will be poor as it encoder would need to degrade the quality by raising the compression level to stay at target. The frame rate may be degraded if raising the compression level is not sufficient.

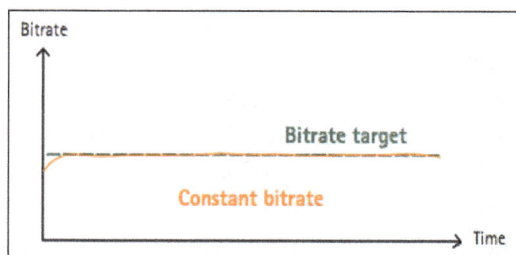

Maximum Bit Rate (MBR)

Maximum bit rate goal is to ensure that you stay under the determined target whatever happens on the video. It uses the same techniques as Constant Bit Rate to ensure that we stay under the set maximum bandwidth. The limit set should be a strict limit which shouldn't be crossed but the bit rate should still stay just under it. It is designed to ensure that we don't lose data because of network limitations and that the video will be streamed whatever happens on the video.

To stay under the maximum bit rate, the bit rate controller changes the video quality and as last option reduce the frame rate.

- Low complexity scene: The bit rate will be the one set as target, there will be no padding so the video bit rate may be under the target sometimes. Video will have full frame rate.

- High complexity scene: The bit rate will still be one set as target but the visual quality will be poor as it encoder would need to degrade the quality by raising the compression level to stay under target. The frame rate may be degraded if raising the compression level is not sufficient.

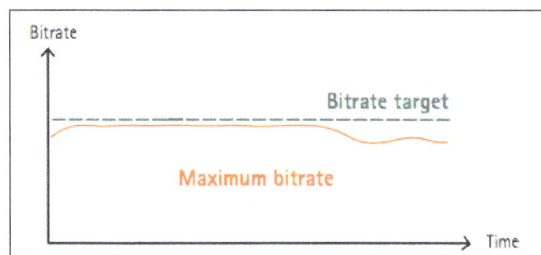

Actual Behavior and Limitations

We will as such focus on two simple examples:

- Low complexity scene (with little/no motion).

- High complexity scene (with medium/big motion).

Variable Bit Rate (VBR)

Axis cameras are compliant with the H.264 level 4.1, which states that the maximum allowed bandwidth s 50 Mbits/s. Due to this limitation the bit rate has to be limited to 50 Mbits/s even in Variable Bit Rate mode to ensure that it will be compliant with other systems and will stay within this bandwidth limit. This may lead to degradation of quality and/or frame rate if the bandwidth needed by the video is more than 50 Mbits/s, yet again to ensure staying compliant with the H.264 level.

- Low complexity scene: The bit rate should be low, but as soon as something occurs then the bit rate will jump to a higher level.

- High complexity scene: The bit rate will be high and may overload the recording device, network or video player.

Parameters that influence this mode:

- Compression level: Sets the quality required on the video, lower compression will provide higher visual quality but higher bit rate, higher compression will provide smaller bit rate but also worse video quality.

- GOV length: Sets the length of the Group Of Pictures (GOP) used for the video. A smaller GOP leads to more frequent I-frames so a higher bit rate, bigger values will result in less I-frames so lower bit rate. Too big GOP will make the seek function on recorded videos more complicated as it needs to refer to an I-frame to render following P-frames of the video, also in case of data loss, visual artifacts will stay longer as only the I-frames are fully updated frames.

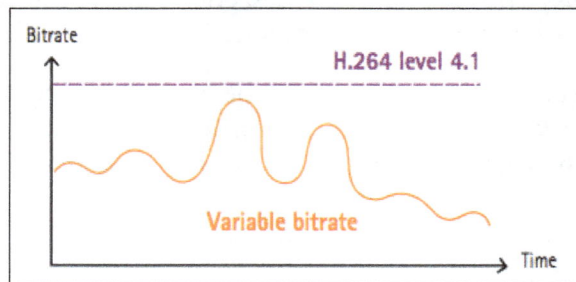

Constant Bit Rate (CBR)

The standard definition of constant is to be understood as in average bit rate, the goal of a bit rate controller is to achieve this target on a short average by changing the quality or frame rate of the video.

This implies that Constant bit rate is never strictly constant in reality, at least from the instantaneous point of view (few frames) which can be critical on a network. The changes are dependent on the scene and the movement, the more complex the scene, the worse visual quality it will be for a same target.

CBR can lead to big bandwidth changes on the short term (few frames), time used by the bit rate controller to adapt the parameters to the new scene.

In the Axis cameras, bit padding is not done, that may lead to having a bit rate under the target:

- Low complexity scene: The bit rate should be low, as soon as something happens, the visual quality will be degraded. It is possible that the target bit rate is not achieved because we can't provide enough data to get to the target bit rate. In case of changes overshoots can occur in the short term (few frames).

- High complexity scene: The visual quality will be degraded and frame rate diminished if needed, this may lead to several (big in amplitude but small in time) overshoots due to frequent changes in the scene.

Parameters that influence this mode:

- Compression level: Sets the best quality possible for the video, lower compression will provide Higher visual quality but higher possible bit rate, higher compression will provide smaller bit rate but also worse video quality and also may lead to lower bit rate than target.

- Target bit rate: Sets the target to achieve (in kbits/s).

- GOV length: Sets the length of the Group Of Pictures (GOP) used for the video. Lower value will lead to smaller GOP, more frequent I-frames so higher bit rate, bigger values will result in less I-frames so lower bit rate. Too big GOP will make the seek function on recorded videos more complicated as it needs to refer to an I-frame to render following P-frames of the video, also in case of data loss, visual artifacts will stay longer as only the I-frames are the fully updated frames.

- Priority: Set the priority of the bit rate controller to none, quality or framerate. These parameters load different margins for possible quality levels and frame skipped at once. Quality option will lead to the best visual quality at anytime, most likely at the cost of frame rate. Framerate option will lead to a possible worse visual quality but ensuring a certain frame rate.

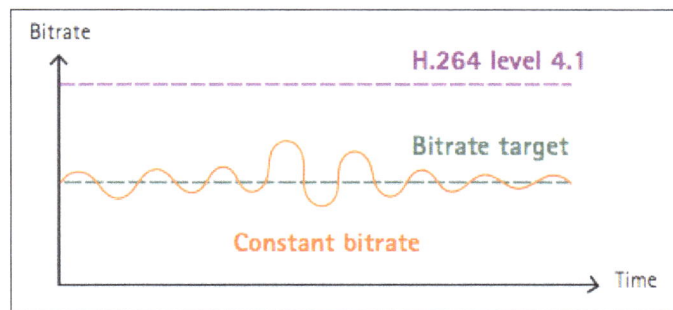

Maximum Bit Rate (MBR)

Maximum Bit Rate was designed so that the bandwidth of a video stays under the selected maximum bit rate and should ensure that both long term and short term bandwidth stays under this limit.

To do so it reacts fast on scene changes to limit spikes on bitrate as much as possible (most of the time only a few frames long) during a change. It drops frames if needed to ensure an overshoot as small as possible on the instantaneous level.

This can lead to the impression of frame rate drop when not reaching the set maximum bit rate, this is due to a mismatch between the average bandwidth displayed and the instantaneous bit rate as this rate controller works on both time scales.

In the Axis cameras, bit padding is not done, that may lead to having a bit rate way under the maximum value desired.

- Low complexity scene: The bit rate should be low, as soon as something happens, the visual quality will be degraded. It is possible that the selected maximum bit rate is not achieved because we can't provide enough data to get close to the maximum bit rate. In case of changes, overshoots in the short term will be very limited.

- High complexity scene: The visual quality will be degraded and frame rate diminished if needed, this may lead to several (small in amplitude and in time) overshoots due to frequent changes in the scene, frame drops will be used to ensure a minimal spike on the bit rate.

Parameters that influence this mode:

- Compression level: Sets the best quality possible for the video, lower compression will provide higher visual quality but higher possible bit rate, higher compression will provide smaller bit rate but also worse video quality and also may lead to lower bit rate than target.

- Target bit rate: Sets the target to achieve as maximum (in kbits/s).

- GOV length: Sets the length of the Group Of Pictures (GOP) used for the video. Lower value will lead to smaller GOP, more frequent I-frames so higher bit rate, bigger values will result in less I-frames so lower bit rate. Too big GOP will make the seek function on recorded videos more complicated as it needs to refer to an I-frame to render following P-frames of the video, also in case of data loss, visual artifacts will stay longer as only the I-frames are fully updated frames.

- Priority: Set the priority of the bit rate controller to none, quality or framerate. These parameters load different margins for possible quality levels and frame skipped at once. Quality option will lead to the best visual quality at anytime, most likely at the cost of frame rate. Framerate option will lead to a possible worse visual quality but ensuring a certain frame rate.

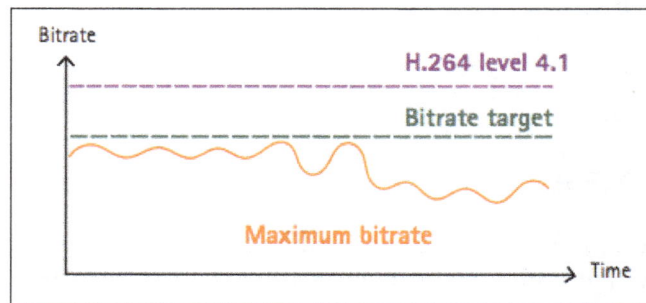

BAUD RATE

In digital communications, symbol rate, also known as baud rate and modulation rate, is the number of symbol changes, waveform changes, or signaling events, across the transmission medium per time unit using a digitally modulated signal or a line code. The symbol rate is measured in baud (Bd) or symbols per second. In the case of a line code, the symbol rate is the pulse rate in pulses per second. Each symbol can represent or convey one or several bits of data. The symbol rate is related to the gross bitrate expressed in bits per second.

Symbols

A symbol may be described as either a pulse in digital baseband transmission or a tone in passband transmission using modems. A symbol is a waveform, a state or a significant condition of the communication channel that persists, for a fixed period of time. A sending device places symbols on the channel at a fixed and known symbol rate, and the receiving device has the job of detecting the sequence of symbols in order to reconstruct the transmitted data. There may be a direct correspondence between a symbol and a small unit of data. For example, each symbol may encode one or several binary digits or 'bits'. The data may also be represented by the transitions between symbols, or even by a sequence of many symbols.

The symbol duration time, also known as unit interval, can be directly measured as the time between transitions by looking into an eye diagram of an oscilloscope. The symbol duration time T_s can be calculated as:

$$T_s = \frac{1}{f_s}$$

where f_s is the symbol rate.

A simple example: A baud rate of 1 kBd = 1,000 Bd is synonymous to a symbol rate of 1,000 symbols per second. In case of a modem, this corresponds to 1,000 tones per second, and in case of a line code, this corresponds to 1,000 pulses per second. The symbol duration time is 1/1,000 second = 1 millisecond.

Relationship to Gross Bitrate

The term baud rate has sometimes incorrectly been used to mean bit rate, since these rates are the same in old modems as well as in the simplest digital communication links using only one bit per symbol, such that binary "0" is represented by one symbol, and binary "1" by another symbol. In more advanced modems and data transmission techniques, a symbol may have more than two states, so it may represent more than one binary digit (a binary digit always represents one of exactly two states). For this reason, the baud rate value will often be lower than the gross bit rate.

Example of use and misuse of "baud rate": It is correct to write "the baud rate of my COM port is 9,600" if we mean that the bit rate is 9,600 bit/s, since there is one bit per symbol in this case. It is not correct to write "the baud rate of Ethernet is 100 megabaud" or "the baud rate of my modem is 56,000" if we mean bit rate.

The difference between baud (or signalling rate) and the data rate (or bit rate) is like a man using a single semaphore flag who can move his arm to a new position once each second, so his signalling rate (baud) is one symbol per second. The flag can be held in one of eight distinct positions: Straight up, 45° left, 90° left, 135° left, straight down (which is the rest state, where he is sending no signal), 135° right, 90° right, and 45° right. Each signal (symbol) carries three bits of information. It takes three binary digits to encode eight states. The data rate is three bits per second. In the Navy, more than one flag pattern and arm can be used at once, so the combinations of these produce many symbols, each conveying several bits, a higher data rate.

If N bits are conveyed per symbol, and the gross bit rate is R, inclusive of channel coding overhead, the symbol rate can be calculated as:

$$f_s = \frac{R}{N}$$

In that case $M = 2^N$ different symbols are used. In a modem, these may be sinewave tones with unique combinations of amplitude, phase and/or frequency. For example, in a 64QAM modem, $M = 64$. In a line code, these may be M different voltage levels.

By taking information per pulse N in bit/pulse to be the base-2-logarithm of the number of distinct messages M that could be sent, Hartley constructed a measure of the gross bitrate R as,

$$R = f_s \log_2(M)$$

where f_s is the baud rate in symbols/second or pulses/second.

Modems for Passband Transmission

Modulation is used in passband filtered channels such as telephone lines, radio channels and other frequency division multiplex (FDM) channels.

In a digital modulation method provided by a modem, each symbol is typically a sine wave tone with a certain frequency, amplitude and phase. Symbol rate, baud rate, is the number of transmitted tones per second.

One symbol can carry one or several bits of information. In voiceband modems for the telephone network, it is common for one symbol to carry up to 7 bits.

Conveying more than one bit per symbol or bit per pulse has advantages. It reduces the time required to send a given quantity of data over a limited bandwidth. A high spectral efficiency in (bit/s)/Hz can be achieved; i.e., a high bit rate in bit/s although the bandwidth in hertz may be low.

The maximum baud rate for a passband for common modulation methods such as QAM, PSK and OFDM is approximately equal to the passband bandwidth.

Voiceband modem examples:

- A V.22bis modem transmits 2400 bit/s using 1200 Bd (1200 symbol/s), where each quadrature amplitude modulation symbol carries two bits of information. The modem can generate $M=2^2=4$ different symbols. It requires a bandwidth of 1200 Hz (equal to the baud rate). The carrier frequency is 1800 Hz, meaning that the lower cut off frequency is $1,800 - 1,200/2 = 1,200$ Hz, and the upper cutoff frequency is $1,800 + 1,200/2 = 2,400$ Hz.

- A V.34 modem may transmit symbols at a baud rate of 3,420 Bd, and each symbol can carry up to ten bits, resulting in a gross bit rate of $3420 \times 10 = 34,200$ bit/s. However, the modem is said to operate at a net bit rate of 33,800 bit/s, excluding physical layer overhead.

Line Codes for Baseband Transmission

In case of a baseband channel such as a telegraph line, a serial cable or a Local Area Network twisted pair cable, data is transferred using line codes; i.e., pulses rather than sinewave tones. In this case, the baud rate is synonymous to the pulse rate in pulses/second.

The maximum baud rate or pulse rate for a base band channel is called the Nyquist rate, and is double the bandwidth (double the cut-off frequency).

The simplest digital communication links (such as individual wires on a motherboard or the RS-232 serial port/COM port) typically have a symbol rate equal to the gross bit rate.

Common communication links such as 10 Mbit/s Ethernet (10Base-T), USB, and FireWire typically have a symbol rate slightly lower than the data bit rate, due to the overhead of extra non-data symbols used for self-synchronizing code and error detection.

J. M. Emile Baudot worked out a five-level code (five bits per character) for telegraphs which was standardized internationally and is commonly called Baudot code.

More than two voltage levels are used in advanced techniques such as FDDI and 100/1,000 Mbit/s Ethernet LANs, and others, to achieve high data rates.

1,000 Mbit/s Ethernet LAN cables use four wire pairs in full duplex (250 Mbit/s per pair in both directions simultaneously), and many bits per symbol to encode their data payloads.

Digital Television and OFDM Example

In digital television transmission the symbol rate calculation is:

symbol rate in symbols per second = (Data rate in bits per second × 204)/(188 × bits per symbol)

The 204 is the number of bytes in a packet including the 16 trailing Reed-Solomon error checking and correction bytes. The 188 is the number of data bytes (187 bytes) plus the leading packet sync byte (0x47).

The bits per symbol is the (modulation's power of 2) × (Forward Error Correction). So for example, in 64-QAM modulation $64 = 2^6$ so the bits per symbol is 6. The Forward Error Correction (FEC) is usually expressed as a fraction; i.e., 1/2, 3/4, etc. In the case of 3/4 FEC, for every 3 bits of data, you are sending out 4 bits, one of which is for error correction.

Example:

given bit rate = 18096263,

Modulation type = 64-QAM,

FEC = 3/4,

then,

$$\text{symbol rate} = \frac{18096263}{6 \cdot \frac{3}{4}} \frac{204}{188} = \frac{18096263}{6} \frac{4}{3} \frac{204}{188} = 4363638$$

In digital terrestrial television (DVB-T, DVB-H and similar techniques) OFDM modulation is used; i.e., multi-carrier modulation. The above symbol rate should then be divided by the number of OFDM sub-carriers in view to achieve the OFDM symbol rate.

Relationship to Chip Rate

Some communication links (such as GPS transmissions, CDMA cell phones, and other spread spectrum links) have a symbol rate much higher than the data rate (they transmit many symbols called chips per data bit). Representing one bit by a chip sequence of many symbols overcomes co-channel interference from other transmitters sharing the same frequency channel, including radio jamming, and is common in military radio and cell phones. Despite the fact that using more bandwidth to carry the same bit rate gives low channel spectral efficiency in (bit/s)/Hz, it allows many simultaneous users, which results in high system spectral efficiency in (bit/s)/Hz per unit of area.

In these systems, the symbol rate of the physically transmitted high-frequency signal rate is called chip rate, which also is the pulse rate of the equivalent base band signal. However, in spread spectrum systems, the term symbol may also be used at a higher layer and refer to one information bit, or a block of information bits that are modulated using for example conventional QAM modulation, before the CDMA spreading code is applied. Using the latter definition, the symbol rate is equal to or lower than the bit rate.

Relationship to Bit Error Rate

The disadvantage of conveying many bits per symbol is that the receiver has to distinguish many signal levels or symbols from each other, which may be difficult and cause bit errors in case of a poor phone line that suffers from low signal-to-noise ratio. In that case, a modem or network adapter may automatically choose a slower and more robust modulation scheme or line code, using fewer bits per symbol, in view to reduce the bit error rate.

An optimal symbol set design takes into account channel bandwidth, desired information rate, noise characteristics of the channel and the receiver, and receiver and decoder complexity.

Modulation

Many data transmission systems operate by the modulation of a carrier signal. For example, in frequency-shift keying (FSK), the frequency of a tone is varied among a small, fixed set of possible values. In a synchronous data transmission system, the tone can only be changed from one frequency to another at regular and well-defined intervals. The presence of one particular frequency during one of these intervals constitutes a symbol. (The concept of symbols does not apply to asynchronous data transmission systems.) In a modulated system, the term modulation rate may be used synonymously with symbol rate.

Binary Modulation

If the carrier signal has only two states, then only one bit of data (i.e., a 0 or 1) can be transmitted in each symbol. The bit rate is in this case equal to the symbol rate. For example, a binary FSK system would allow the carrier to have one of two frequencies, one representing a 0 and the other a 1. A more practical scheme is differential binary phase-shift keying, in which the carrier remains at the same frequency, but can be in one of two phases. During each symbol, the phase either remains the same, encoding a 0, or jumps by 180°, encoding a 1. Again, only one bit of data (i.e., a 0 or 1) is transmitted by each symbol. This is an example of data being encoded in the transitions between symbols (the change in phase), rather than the symbols themselves (the actual phase). (The reason for this in phase-shift keying is that it is impractical to know the reference phase of the transmitter.)

N-ary Modulation, *N* greater than 2

By increasing the number of states that the carrier signal can take, the number of bits encoded in each symbol can be greater than one. The bit rate can then be greater than the symbol rate. For example, a differential phase-shift keying system might allow four possible jumps in phase between symbols. Then two bits could be encoded at each symbol interval, achieving a data rate of double the symbol rate. In a more complex scheme such as 16-QAM, four bits of data are transmitted in each symbol, resulting in a bit rate of four times the symbol rate.

Not Power of 2

Although it is common to choose the number of symbols to be a power of 2 and send an integer number of bits per baud, this is not required. Line codes such as bipolar encoding and MLT-3 use three carrier states to encode one bit per baud while maintaining DC balance.

The 4B3T line code uses three 3-ary modulated bits to transmit four data bits, a rate of 1.33 bits per baud.

Data Rate Versus Error Rate

Modulating a carrier increases the frequency range, or bandwidth, it occupies. Transmission channels are generally limited in the bandwidth they can carry. The bandwidth depends on the symbol (modulation) rate (not directly on the bit rate). As the bit rate is the product of the symbol rate and the number of bits encoded in each symbol, it is clearly advantageous to increase the latter if the former is fixed. However, for each additional bit encoded in a symbol, the constellation of symbols (the number of states of the carrier) doubles in size. This makes the states less distinct from one another which in turn makes it more difficult for the receiver to detect the symbol correctly in the presence of disturbances on the channel.

The history of modems is the attempt at increasing the bit rate over a fixed bandwidth (and therefore a fixed maximum symbol rate), leading to increasing bits per symbol. For example, the V.29 specifies 4 bits per symbol, at a symbol rate of 2,400 baud, giving an effective bit rate of 9,600 bits per second.

The history of spread spectrum goes in the opposite direction, leading to fewer and fewer data bits per symbol in order to spread the bandwidth. In the case of GPS, we have a data rate of 50 bit/s and a symbol rate of 1.023 Mchips/s. If each chip is considered a symbol, each symbol contains far less than one bit (50 bit/s/1,023 ksymbols/s $\approx$ 0.000,05 bits/symbol).

The complete collection of M possible symbols over a particular channel is called a M-ary modulation scheme. Most modulation schemes transmit some integer number of bits per symbol b, requiring the complete collection to contain $M = 2^b$ different symbols. Most popular modulation schemes can be described by showing each point on a constellation diagram, although a few modulation schemes (such as MFSK, DTMF, pulse-position modulation, spread spectrum modulation) require a different description.

Significant Condition

In telecommunication, concerning the modulation of a carrier, a significant condition is one of the signal's parameters chosen to represent information.

A significant condition could be an electric current (voltage, or power level), an optical power level, a phase value, or a particular frequency or wavelength. The duration of a significant condition is the time interval between successive significant instants. A change from one significant condition to another is called a signal transition. Information can be transmitted either during the given time interval, or encoded as the presence or absence of a change in the received signal.

Significant conditions are recognized by an appropriate device called a receiver, demodulator, or decoder. The decoder translates the actual signal received into its intended logical value such as a binary digit (0 or 1), an alphabetic character, a mark, or a space. Each significant instant is determined when the appropriate device assumes a condition or state usable for performing a specific function, such as recording, processing, or gating.

Baud

In telecommunication and electronics, baud is a common measure of symbol rate, one of the components that determine the speed of communication over a data channel.

It is the unit for symbol rate or modulation rate in symbols per second or pulses per second. It is the number of distinct symbol changes (signaling events) made to the transmission medium per second in a digitally modulated signal or a bd rate line code.

Baud is related to, but not equivalent to, gross bit rate, which can be expressed as bits per second. If there are only two symbols in the system (typically 0 and 1), then baud and bits per second (bps) are equivalent.

Naming

The baud unit is named after Émile Baudot, the inventor of the Baudot code for telegraphy, and is represented in accordance with the rules for SI units. That is, the first letter of its symbol is uppercase (Bd), but when the unit is spelled out, it should be written in lowercase (baud) except when it begins a sentence. It was defined by the CCITT (now the ITU) in November 1926. The earlier standard had been the number of words per minute. One baud was equal to one pulse per second, a more robust measure as word length can vary.

The symbol duration time, also known as unit interval, can be directly measured as the time between transitions by looking at an eye diagram of the signal on an oscilloscope. The symbol duration time T_s can be calculated as,

$$T_s = \frac{1}{f_s},$$

where f_s is the symbol rate. There is also a chance of miscommunication which leads to ambiguity.

Example: Communication at the baud rate 1000 Bd means communication by means of sending 1000 symbols per second. In the case of a modem, this corresponds to 1000 tones per second; similarly, in the case of a line code, this corresponds to 1000 pulses per second. The symbol duration time is 1/1000 second (that is, 1 millisecond).

In digital systems (i.e., using discrete/discontinuous values) with binary code, 1 Bd = 1 bit/s. By contrast, non-digital (or analog) systems use a continuous range of values to represent information and in these systems the exact informational size of 1 Bd varies.

The baud is scaled using standard metric prefixes, so that for example:

- 1 kBd (kilobaud) = 1000 Bd,

- 1 MBd (megabaud) = 1000 kBd,

- 1 GBd (gigabaud) = 1000 MBd.

Relationship to Gross Bit Rate

The symbol rate is related to gross bit rate expressed in bit/s. The term baud has sometimes incorrectly been used to mean bit rate, since these rates are the same in old modems as well as in the simplest digital communication links using only one bit per symbol, such that binary digit "0" is represented by one symbol, and binary digit "1" by another symbol. In more advanced modems and data transmission techniques, a symbol may have more than two states, so it may represent more than one bit. A bit (binary digit) always represents one of two states.

If N bits are conveyed per symbol, and the gross bit rate is R, inclusive of channel coding overhead, the symbol rate f_s can be calculated as,

$$f_s = \frac{R}{N}.$$

By taking information per pulse N in bit/pulse to be the base-2-logarithm of the number of distinct messages M that could be sent, Hartley constructed a measure of the gross bitrate R as,

$$R = f_s N,$$

where $N = \log_2(M)$.

In that case $M = 2^N$, different symbols are used. In a modem, these may be time-limited sinewave tones with unique combinations of amplitude, phase and/or frequency. For example, in a 64QAM modem, $M = 64$, and so the bit rate is $N = \log_2(64) = 6$ times the baud rate. In a line code, these may be M different voltage levels.

The ratio is not necessarily even an integer; in 4B3T coding, the bit rate is 4/3 of the baud rate. (A typical basic rate interface with a 160 kbit/s raw data rate operates at 120 kBd.)

Codes with many symbols, and thus a bit rate higher than the symbol rate, are most useful on channels such as telephone lines with a limited bandwidth but a high signal-to-noise ratio within that bandwidth. In other applications, the bit rate is less than the symbol rate. Eight-to-fourteen modulation as used on audio CDs has bit rate 8/14 of the baud rate.

ERROR DETECTION AND CORRECTION CODE

In information theory and coding theory with applications in computer science and telecommunication, error detection and correction or error control are techniques that enable reliable delivery of digital data over unreliable communication channels. Many communication channels are subject to channel noise, and thus errors may be introduced during transmission from the source to a receiver. Error detection techniques allow detecting such errors, while error correction enables reconstruction of the original data in many cases.

All error-detection and correction schemes add some redundancy (i.e., some extra data) to a message, which receivers can use to check consistency of the delivered message, and to recover data that has been determined to be corrupted. Error-detection and correction schemes can be either systematic or non-systematic. In a systematic scheme, the transmitter sends the original data, and attaches a fixed number of check bits (or parity data), which are derived from the data bits by some deterministic algorithm. If only error detection is required, a receiver can simply apply the same algorithm to the received data bits and compare its output with the received check bits; if the values do not match, an error has occurred at some point during the transmission. In a system that uses a non-systematic code, the original message is transformed into an encoded message carrying the same information and that has at least as many bits as the original message.

Good error control performance requires the scheme to be selected based on the characteristics of the communication channel. Common channel models include memoryless models where errors occur randomly and with a certain probability, and dynamic models where errors occur primarily in bursts. Consequently, error-detecting and correcting codes can be generally distinguished between random-error-detecting/correcting and burst-error-detecting/correcting. Some codes can also be suitable for a mixture of random errors and burst errors.

If the channel characteristics cannot be determined, or are highly variable, an error-detection scheme may be combined with a system for retransmissions of erroneous data. This is known as automatic repeat request (ARQ), and is most notably used in the Internet. An alternate approach for error control is hybrid automatic repeat request (HARQ), which is a combination of ARQ and error-correction coding.

Types

Error correction may generally be realized in two different ways:

- Automatic repeat request (ARQ) (sometimes also referred to as backward error correction): This is an error control technique whereby an error detection scheme is combined with requests for retransmission of corrupted data. Every block of data received is checked using the error detection code used, and if the check fails, retransmission of the data is requested – this may be done repeatedly, until the data can be verified.

- Forward error correction (FEC): The sender encodes the data using an error-correcting code (ECC) prior to transmission. The additional information (redundancy) added by the code is used by the receiver to recover the original data in the case of corruption.

ARQ and FEC may be combined such that minor errors are corrected without retransmission, and major errors are corrected via a request for retransmission: this is called hybrid automatic repeat request (HARQ).

Error Detection Schemes

Error detection is most commonly realized using a suitable hash function (or specifically, a checksum, cyclic redundancy check or other algorithm). A hash function adds a fixed-length tag to a message, which enables receivers to verify the delivered message by recomputing the tag and comparing it with the one provided.

There exists a vast variety of different hash function designs. However, some are of particularly widespread use because of either their simplicity or their suitability for detecting certain kinds of errors (e.g., the cyclic redundancy check's performance in detecting burst errors).

Minimum Distance Coding

A random-error-correcting code based on minimum distance coding can provide a strict guarantee on the number of detectable errors, but it may not protect against a preimage attack.

Repetition Codes

A repetition code is a special case of error-correcting code: although rather inefficient, a repetition code is suitable in some applications of error correction and detection due to its simplicity.

A repetition code is a coding scheme that repeats the bits across a channel to achieve error-free communication. Given a stream of data to be transmitted, the data are divided into blocks of bits. Each block is transmitted some predetermined number of times. For example, to send the bit pattern "1011", the four-bit block can be repeated three times, thus producing "1011 1011 1011". However, if this twelve-bit pattern was received as "1010 1011 1011" – where the first block is unlike the other two – it can be determined that an error has occurred.

A repetition code is very inefficient, and can be susceptible to problems if the error occurs in exactly the same place for each group (e.g., "1010 1010 1010" in the previous example would be detected as correct). The advantage of repetition codes is that they are extremely simple, and are in fact used in some transmissions of numbers stations.

Parity bits

A parity bit is a bit that is added to a group of source bits to ensure that the number of set bits (i.e., bits with value 1) in the outcome is even or odd. It is a very simple scheme that can be used to detect single or any other odd number (i.e., three, five, etc.) of errors in the output. An even number of flipped bits will make the parity bit appear correct even though the data is erroneous.

Extensions and variations on the parity bit mechanism are horizontal redundancy checks, vertical redundancy checks, and "double," "dual," or "diagonal" parity (used in RAID-DP).

Checksums

A checksum of a message is a modular arithmetic sum of message code words of a fixed word length (e.g., byte values). The sum may be negated by means of a ones'-complement operation prior to transmission to detect errors resulting in all-zero messages.

Checksum schemes include parity bits, check digits, and longitudinal redundancy checks. Some checksum schemes, such as the Damm algorithm, the Luhn algorithm, and the Verhoeff algorithm, are specifically designed to detect errors commonly introduced by humans in writing down or remembering identification numbers.

Cyclic Redundancy Checks (CRCs)

A cyclic redundancy check (CRC) is a non-secure hash function designed to detect accidental changes to digital data in computer networks; as a result, it is not suitable for detecting maliciously introduced errors. It is characterized by specification of what is called a generator polynomial, which is used as the divisor in a polynomial long division over a finite field, taking the input data as the dividend, such that the remainder becomes the result.

A cyclic code has favorable properties that make it well suited for detecting burst errors. CRCs are particularly easy to implement in hardware, and are therefore commonly used in digital networks and storage devices such as hard disk drives.

Even parity is a special case of a cyclic redundancy check, where the single-bit CRC is generated by the divisor $x + 1$.

Cryptographic Hash Functions

The output of a cryptographic hash function, also known as a message digest, can provide strong assurances about data integrity, whether changes of the data are accidental (e.g., due to transmission errors) or maliciously introduced. Any modification to the data will likely be detected through a mismatching hash value. Furthermore, given some hash value, it is infeasible to find some input data (other than the one given) that will yield the same hash value. If an attacker can change not only the message but also the hash value, then a keyed hash or message authentication code (MAC) can be used for additional security. Without knowing the key, it is not possible for the attacker to easily or conveniently calculate the correct keyed hash value for a modified message.

Error-correcting Codes

Any error-correcting code can be used for error detection. A code with minimum Hamming distance, d, can detect up to $d - 1$ errors in a code word. Using minimum-distance-based error-correcting codes for error detection can be suitable if a strict limit on the minimum number of errors to be detected is desired.

Codes with minimum Hamming distance $d = 2$ are degenerate cases of error-correcting codes, and can be used to detect single errors. The parity bit is an example of a single-error-detecting code.

Error Correction

Automatic Repeat Request (ARQ)

Automatic Repeat reQuest (ARQ) is an error control method for data transmission that makes use of error-detection codes, acknowledgment and/or negative acknowledgment messages, and timeouts to achieve reliable data transmission. An acknowledgment is a message sent by the receiver to indicate that it has correctly received a data frame.

Usually, when the transmitter does not receive the acknowledgment before the timeout occurs (i.e., within a reasonable amount of time after sending the data frame), it retransmits the frame until it is either correctly received or the error persists beyond a predetermined number of retransmissions.

Three types of ARQ protocols are Stop-and-wait ARQ, Go-Back-N ARQ, and Selective Repeat ARQ.

ARQ is appropriate if the communication channel has varying or unknown capacity, such as is the case on the Internet. However, ARQ requires the availability of a back channel, results in possibly increased latency due to retransmissions, and requires the maintenance of buffers and timers for retransmissions, which in the case of network congestion can put a strain on the server and overall network capacity.

For example, ARQ is used on shortwave radio data links in the form of ARQ-E, or combined with multiplexing as ARQ-M.

Error-correcting Code

An error-correcting code (ECC) or forward error correction (FEC) code is a process of adding redundant data, or parity data, to a message, such that it can be recovered by a receiver even when a number of errors (up to the capability of the code being used) were introduced, either during the process of transmission, or on storage. Since the receiver does not have to ask the sender for retransmission of the data, a backchannel is not required in forward error correction, and it is therefore suitable for simplex communication such as broadcasting. Error-correcting codes are frequently used in lower-layer communication, as well as for reliable storage in media such as CDs, DVDs, hard disks, and RAM.

Error-correcting codes are usually distinguished between convolutional codes and block codes:

- Convolutional codes are processed on a bit-by-bit basis. They are particularly suitable for implementation in hardware, and the Viterbi decoder allows optimal decoding.

- Block codes are processed on a block-by-block basis. Early examples of block codes are repetition codes, Hamming codes and multidimensional parity-check codes. They were followed by a number of efficient codes, Reed–Solomon codes being the most notable due to their current widespread use. Turbo codes and low-density parity-check codes (LDPC) are relatively new constructions that can provide almost optimal efficiency.

Shannon's theorem is an important theorem in forward error correction, and describes the maximum information rate at which reliable communication is possible over a channel that has a certain error probability or signal-to-noise ratio (SNR). This strict upper limit is expressed in terms of the channel capacity. More specifically, the theorem says that there exist codes such that with increasing encoding length the probability of error on a discrete memoryless channel can be made arbitrarily small, provided that the code rate is smaller than the channel capacity. The code rate is defined as the fraction k/n of k source symbols and n encoded symbols.

The actual maximum code rate allowed depends on the error-correcting code used, and may be lower. This is because Shannon's proof was only of existential nature, and did not show how to construct codes which are both optimal and have efficient encoding and decoding algorithms.

Hybrid Schemes

Hybrid ARQ is a combination of ARQ and forward error correction. There are two basic approaches:

- Messages are always transmitted with FEC parity data (and error-detection redundancy). A receiver decodes a message using the parity information, and requests retransmission using ARQ only if the parity data was not sufficient for successful decoding (identified through a failed integrity check).

- Messages are transmitted without parity data (only with error-detection information). If a receiver detects an error, it requests FEC information from the transmitter using ARQ, and uses it to reconstruct the original message.

The latter approach is particularly attractive on an erasure channel when using a rateless erasure code.

Applications

Applications that require low latency (such as telephone conversations) cannot use Automatic Repeat reQuest (ARQ); they must use forward error correction (FEC). By the time an ARQ system discovers an error and re-transmits it, the re-sent data will arrive too late to be any good.

Applications where the transmitter immediately forgets the information as soon as it is sent (such as most television cameras) cannot use ARQ; they must use FEC because when an error occurs, the original data is no longer available. (This is also why FEC is used in data storage systems such as RAID and distributed data store).

Applications that use ARQ must have a return channel; applications having no return channel cannot use ARQ. Applications that require extremely low error rates (such as digital money transfers) must use ARQ. Reliability and inspection engineering also make use of the theory of error-correcting codes.

Internet

In a typical TCP/IP stack, error control is performed at multiple levels:

- Each Ethernet frame carries a CRC-32 checksum. Frames received with incorrect checksums are discarded by the receiver hardware.

- The IPv4 header contains a checksum protecting the contents of the header. Packets with mismatching checksums are dropped within the network or at the receiver.

- The checksum was omitted from the IPv6 header in order to minimize processing costs in network routing and because current link layer technology is assumed to provide sufficient error detection.

- UDP has an optional checksum covering the payload and addressing information from the UDP and IP headers. Packets with incorrect checksums are discarded by the operating system network stack. The checksum is optional under IPv4, only, because the Data-Link layer checksum may already provide the desired level of error protection.

- TCP provides a checksum for protecting the payload and addressing information from the TCP and IP headers. Packets with incorrect checksums are discarded within the network stack, and eventually get retransmitted using ARQ, either explicitly (such as through triple-ack) or implicitly due to a timeout.

Deep-space Telecommunications

Development of error-correction codes was tightly coupled with the history of deep-space missions due to the extreme dilution of signal power over interplanetary distances, and the limited power availability aboard space probes. Whereas early missions sent their data uncoded, starting from 1968, digital error correction was implemented in the form of (sub-optimally decoded) convolutional codes and Reed–Muller codes. The Reed–Muller code was well suited to the noise the spacecraft was subject to (approximately matching a bell curve), and was implemented at the Mariner spacecraft for missions between 1969 and 1977.

The Voyager 1 and Voyager 2 missions, which started in 1977, were designed to deliver color imaging amongst scientific information of Jupiter and Saturn. This resulted in increased coding requirements, and thus, the spacecraft were supported by (optimally Viterbi-decoded) convolutional codes that could be concatenated with an outer Golay (24,12,8) code.

The Voyager 2 craft additionally supported an implementation of a Reed–Solomon code: the concatenated Reed–Solomon–Viterbi (RSV) code allowed for very powerful error correction, and enabled the spacecraft's extended journey to Uranus and Neptune. Both crafts used V2 RSV coding due to ECC system upgrades after 1989.

The CCSDS currently recommends usage of error correction codes with performance similar to the Voyager 2 RSV code as a minimum. Concatenated codes are increasingly falling out of favor with space missions, and are replaced by more powerful codes such as Turbo codes or LDPC codes.

The different kinds of deep space and orbital missions that are conducted suggest that trying to find a "one size fits all" error correction system will be an ongoing problem for some time to come. For missions close to Earth, the nature of the noise in the communication channel is different from that which a spacecraft on an interplanetary mission experiences. Additionally, as a spacecraft increases its distance from Earth, the problem of correcting for noise gets bigger.

Satellite Broadcasting

The demand for satellite transponder bandwidth continues to grow, fueled by the desire to deliver television (including new channels and high-definition television) and IP data. Transponder availability and bandwidth constraints have limited this growth, because transponder capacity is determined by the selected modulation scheme and forward error correction (FEC) rate.

- QPSK coupled with traditional Reed Solomon and Viterbi codes have been used for nearly 20 years for the delivery of digital satellite TV.

- Higher order modulation schemes such as 8PSK, 16QAM and 32QAM have enabled the satellite industry to increase transponder efficiency by several orders of magnitude.

- This increase in the information rate in a transponder comes at the expense of an increase in the carrier power to meet the threshold requirement for existing antennas.

- Tests conducted using the latest chipsets demonstrate that the performance achieved by using Turbo Codes may be even lower than the 0.8 dB figure assumed in early designs.

Data Storage

Error detection and correction codes are often used to improve the reliability of data storage media. A "parity track" was present on the first magnetic tape data storage in 1951. The "Optimal Rectangular Code" used in group coded recording tapes not only detects but also corrects single-bit errors. Some file formats, particularly archive formats, include a checksum (most often CRC32) to detect corruption and truncation and can employ redundancy and/or parity files to recover portions of corrupted data. Reed Solomon codes are used in compact discs to correct errors caused by scratches.

Modern hard drives use CRC codes to detect and Reed–Solomon codes to correct minor errors in sector reads, and to recover data from sectors that have "gone bad" and store that data in the spare sectors. RAID systems use a variety of error correction techniques to correct errors when a hard drive completely fails. Filesystems such as ZFS or Btrfs, as well as some RAID implementations, support data scrubbing and resilvering, which allows bad blocks to be detected and (hopefully) recovered before they are used. The recovered data may be re-written to exactly the same physical location, to spare blocks elsewhere on the same piece of hardware, or the data may be rewritten onto replacement hardware.

Error-correcting Memory

DRAM memory may provide stronger protection against soft errors by relying on error correcting codes. Such error-correcting memory, known as *ECC* or *EDAC-protected* memory, is particularly desirable for mission-critical applications, such as scientific computing, financial, medical, etc. as well as deep-space applications due to the increased radiation in space.

Error-correcting memory controllers traditionally use Hamming codes, although some use triple modular redundancy.

Interleaving allows distributing the effect of a single cosmic ray potentially upsetting multiple physically neighboring bits across multiple words by associating neighboring bits to different words. As long as a single event upset (SEU) does not exceed the error threshold (e.g., a single error) in any particular word between accesses, it can be corrected (e.g., by a single-bit error correcting code), and the illusion of an error-free memory system may be maintained.

In addition to hardware providing features required for ECC memory to operate, operating systems usually contain related reporting facilities that are used to provide notifications when soft errors are transparently recovered. An increasing rate of soft errors might indicate that a DIMM module needs replacing, and such feedback information would not be easily available without the related reporting capabilities. One example is the Linux kernel's *EDAC* subsystem (previously known as bluesmoke), which collects the data from error-checking-enabled components inside a computer system; beside collecting and reporting back the events related to ECC memory, it also supports other checksumming errors, including those detected on the PCI bus.

A few systems also support memory scrubbing.

Error Correction Code

In computing, telecommunication, information theory, and coding theory, an error correction code, sometimes error correcting code, (ECC) is used for controlling errors in data over unreliable or noisy communication channels. The central idea is the sender encodes the message with a redundant in the form of an ECC. The American mathematician Richard Hamming pioneered this field in the 1940s and invented the first error-correcting code in 1950: the Hamming (7,4) code. The redundancy allows the receiver to detect a limited number of errors that may occur anywhere in the message, and often to correct these errors without retransmission. ECC gives the receiver the ability to correct errors without needing a reverse channel to request retransmission of data, but at the cost of a fixed, higher forward channel bandwidth. ECC is therefore applied in situations where retransmissions are costly or impossible, such as one-way communication links and when transmitting to multiple receivers in multicast. For example, in the case of a satellite orbiting around Uranus, a retransmission because of decoding errors can create a delay of 5 hours. ECC information is usually added to mass storage devices to enable recovery of corrupted data, is widely used in modems, and is used on systems where the primary memory is ECC memory.

ECC processing in a receiver may be applied to a digital bit stream or in the demodulation of a digitally modulated carrier. For the latter, ECC is an integral part of the initial analog-to-digital conversion in the receiver. The Viterbi decoder implements a soft-decision algorithm to demodulate digital data from an analog signal corrupted by noise. Many ECC encoders/decoders can also generate a bit-error rate (BER) signal which can be used as feedback to fine-tune the analog receiving electronics.

The maximum fractions of errors or of missing bits that can be corrected is determined by the design of the ECC code, so different error correcting codes are suitable for different conditions. In general, a stronger code induces more redundancy that needs to be transmitted using the available bandwidth, which reduces the effective bit-rate while improving the received effective signal-to-noise ratio. The noisy-channel coding theorem of Claude Shannon answers the question of how much bandwidth is left for data communication while using the most efficient code that turns the decoding error probability to zero. This establishes bounds on the theoretical maximum information transfer rate of a channel with some given base noise level. However, the proof is not constructive, and hence gives no insight of how to build a capacity achieving code. After years of research, some advanced ECC systems nowadays come very close to the theoretical maximum.

Working of ECC

ECC is accomplished by adding redundancy to the transmitted information using an algorithm. A redundant bit may be a complex function of many original information bits. The original information may or may not appear literally in the encoded output; codes that include the unmodified input in the output are systematic, while those that do not are non-systematic.

A simplistic example of ECC is to transmit each data bit 3 times, which is known as a repetition code.

Triplet received	Interpreted as
000	0 (error free)
001	0
010	0
100	0
111	1 (error free)
110	1
101	1
011	1

This allows an error in any one of the three samples to be corrected by "majority vote" or "democratic voting". The correcting ability of this ECC is:

- Up to 1 bit of triplet in error,

- Up to 2 bits of triplet omitted.

Though simple to implement and widely used, this triple modular redundancy is a relatively inefficient ECC. Better ECC codes typically examine the last several dozen, or even the last several hundred, previously received bits to determine how to decode the current small handful of bits (typically in groups of 2 to 8 bits).

Averaging Noise to Reduce Errors

ECC could be said to work by "averaging noise"; since each data bit affects many transmitted symbols, the corruption of some symbols by noise usually allows the original user data to be extracted from the other, uncorrupted received symbols that also depend on the same user data.

- Because of this "risk-pooling" effect, digital communication systems that use ECC tend to work well above a certain minimum signal-to-noise ratio and not at all below it.

- This all-or-nothing tendency – the cliff effect – becomes more pronounced as stronger codes are used that more closely approach the theoretical Shannon limit.

- Interleaving ECC coded data can reduce the all or nothing properties of transmitted ECC codes when the channel errors tend to occur in bursts. However, this method has limits; it is best used on narrowband data.

Most telecommunication systems use a fixed channel code designed to tolerate the expected worst-case bit error rate, and then fail to work at all if the bit error rate is ever worse. However, some systems adapt to the given channel error conditions: some instances of hybrid automatic repeat-request use a fixed ECC method as long as the ECC can handle the error rate, then switch to ARQ when the error rate gets too high; adaptive modulation and coding uses a variety of ECC rates, adding more error-correction bits per packet when there are higher error rates in the channel, or taking them out when they are not needed.

Types of ECC

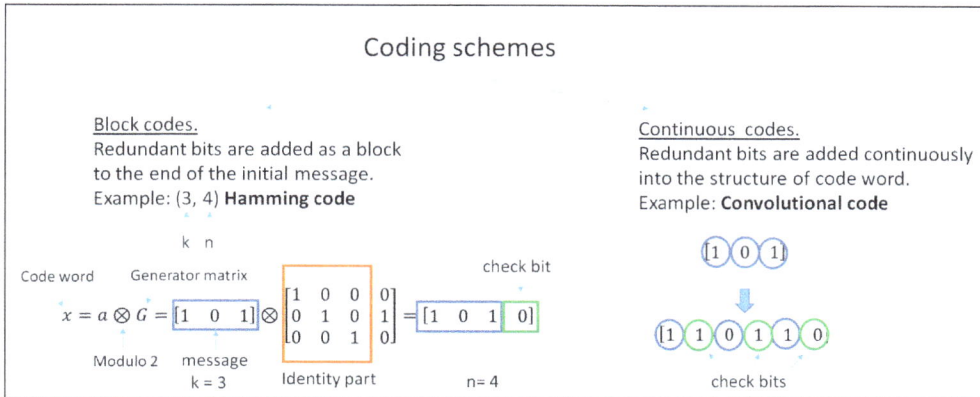

A short classification of the error correction codes.

The two main categories of ECC codes are block codes and convolutional codes:

- Block codes work on fixed-size blocks (packets) of bits or symbols of predetermined size. Practical block codes can generally be hard-decoded in polynomial time to their block length.

- Convolutional codes work on bit or symbol streams of arbitrary length. They are most often soft decoded with the Viterbi algorithm, though other algorithms are sometimes used. Viterbi decoding allows asymptotically optimal decoding efficiency with increasing constraint length of the convolutional code, but at the expense of exponentially increasing complexity. A convolutional code that is terminated is also a 'block code' in that it encodes a block of input data, but the block size of a convolutional code is generally arbitrary, while block codes have a fixed size dictated by their algebraic characteristics. Types of termination for convolutional codes include "tail-biting" and "bit-flushing".

There are many types of block codes, but among the classical ones the most notable is Reed-Solomon coding because of its widespread use in compact discs, DVDs, and hard disk drives. Other examples of classical block codes include Golay, BCH, Multidimensional parity, and Hamming codes.

Hamming ECC is commonly used to correct NAND flash memory errors. This provides single-bit error correction and 2-bit error detection. Hamming codes are only suitable for more reliable single-level cell (SLC) NAND. Denser multi-level cell (MLC) NAND requires stronger multi-bit correcting ECC such as BCH or Reed–Solomon. NOR Flash typically does not use any error correction.

Classical block codes are usually decoded using hard-decision algorithms, which means that for every input and output signal a hard decision is made whether it corresponds to a one or a zero bit. In contrast, convolutional codes are typically decoded using soft-decision algorithms like the Viterbi, MAP or BCJR algorithms, which process (discretized) analog signals, and which allow for much higher error-correction performance than hard-decision decoding.

Nearly all classical block codes apply the algebraic properties of finite fields. Hence classical block codes are often referred to as algebraic codes.

In contrast to classical block codes that often specify an error-detecting or error-correcting ability, many modern block codes such as LDPC codes lack such guarantees. Instead, modern codes are evaluated in terms of their bit error rates.

Most forward error correction codes correct only bit-flips, but not bit-insertions or bit-deletions. In this setting, the Hamming distance is the appropriate way to measure the bit error rate. A few forward error correction codes are designed to correct bit-insertions and bit-deletions, such as Marker Codes and Watermark Codes. The Levenshtein distance is a more appropriate way to measure the bit error rate when using such codes.

Code-rate and the Tradeoff between Reliability and Data Rate

The fundamental principle of ECC is to add redundant bits in order to help the decoder to find out the true message that was encoded by the transmitter. The code-rate of a given ECC system is defined as the rate between the number of information bits and the total number of bits (i.e. information plus redundancy bits) in a given communication package. The code-rate is hence a real number. A low code-rate close to zero implies a strong code that uses many redundant bits to achieve a good performance, while a large code-rate close to 1 implies a weak code.

The redundant bits that protect the information have to be transferred using the same communication resources that they are trying to protect. This causes a fundamental tradeoff between reliability and data rate. In one extreme, a strong code (with low code-rate) can induce an important increase in the receiver SNR decreasing the bit error rate, at the cost of reducing the effective data rate. On the other extreme, not using any ECC (i.e. a code-rate equal to 1) uses the full channel for information transfer purposes, at the cost of leaving the bits without any additional protection.

One interesting question is the following: how efficient in terms of information transfer can be an ECC that has a negligible decoding error rate? This question was answered by Claude Shannon with his second theorem, which says that the channel capacity is the maximum bit rate achievable by any ECC whose error rate tends to zero: His proof relies on Gaussian random coding, which is not suitable of real-world applications. This upper bound given by Shannon's work set up a long journey in designing ECCs that can go close to the ultimate performance boundary. Various codes today can attain almost the Shannon limit. However, capacity achieving ECCs are usually extremely complex to implement.

The most popular ECCs have a trade-off between performance and computational complexity. Usually, their parameters give a range of possible code rates, which can be optimized depending on the scenario. Usually, this optimization is done in order to achieve a low decoding error probability while minimizing the impact to the data rate. Another criterion for optimizing the code rate is to balance low error rate and retransmissions number in order to the energy cost of the communication.

Concatenated ECC Codes for Improved Performance

Classical (algebraic) block codes and convolutional codes are frequently combined in concatenated coding schemes in which a short constraint-length Viterbi-decoded convolutional code does most of the work and a block code (usually Reed-Solomon) with larger symbol size and block length "mops up" any errors made by the convolutional decoder. Single pass decoding with this family of error correction codes can yield very low error rates, but for long range transmission conditions (like deep space) iterative decoding is recommended.

Concatenated codes have been standard practice in satellite and deep space communications since Voyager 2 first used the technique in its 1986 encounter with Uranus. The Galileo craft used iterative concatenated codes to compensate for the very high error rate conditions caused by having a failed antenna.

Low-density Parity-check (LDPC)

Low-density parity-check (LDPC) codes are a class of highly efficient linear block codes made from many single parity check (SPC) codes. They can provide performance very close to the channel capacity (the theoretical maximum) using an iterated soft-decision decoding approach, at linear time complexity in terms of their block length. Practical implementations rely heavily on decoding the constituent SPC codes in parallel.

LDPC codes were first introduced by Robert G. Gallager in his PhD thesis in 1960, but due to the computational effort in implementing encoder and decoder and the introduction of Reed–Solomon codes, they were mostly ignored until the 1990s.

LDPC codes are now used in many recent high-speed communication standards, such as DVB-S2, WiMAX (IEEE 802.16e standard for microwave communications), High-Speed Wireless LAN (IEEE 802.11n), 10GBase-T Ethernet (802.3an) and G.hn/G.9960 (ITU-T Standard for networking over power lines, phone lines and coaxial cable). Other LDPC codes are standardized for wireless communication standards within 3GPP MBMS).

Turbo codes

Turbo coding is an iterated soft-decoding scheme that combines two or more relatively simple convolutional codes and an interleaver to produce a block code that can perform to within a fraction of a decibel of the Shannon limit. Predating LDPC codes in terms of practical application, they now provide similar performance.

One of the earliest commercial applications of turbo coding was the CDMA2000 1x (TIA IS-2000) digital cellular technology developed by Qualcomm and sold by Verizon Wireless, Sprint, and other carriers. It is also used for the evolution of CDMA2000 1x specifically for Internet access, 1xEV-DO (TIA IS-856). Like 1x, EV-DO was developed by Qualcomm, and is sold by Verizon Wireless, Sprint, and other carriers (Verizon's marketing name for 1xEV-DO is Broadband Access, Sprint's consumer and business marketing names for 1xEV-DO are Power Vision and Mobile Broadband, respectively).

Local Decoding and Testing of Codes

Sometimes it is only necessary to decode single bits of the message, or to check whether a given signal is a codeword, and do so without looking at the entire signal. This can make sense in a streaming setting, where codewords are too large to be classically decoded fast enough and where only a few bits of the message are of interest for now. Also such codes have become an important tool in computational complexity theory, e.g., for the design of probabilistically checkable proofs.

Locally decodable codes are error-correcting codes for which single bits of the message can be probabilistically recovered by only looking at a small (say constant) number of positions of a codeword, even after the codeword has been corrupted at some constant fraction of positions. Locally testable codes are error-correcting codes for which it can be checked probabilistically whether a signal is close to a codeword by only looking at a small number of positions of the signal.

Interleaving

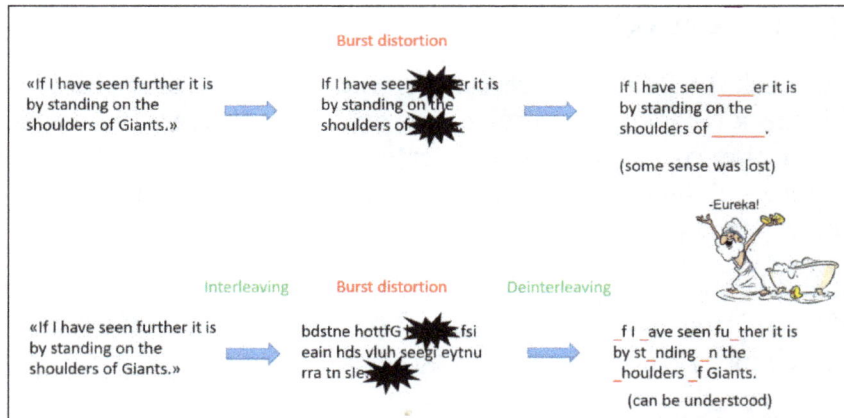

A short illustration of interleaving idea.

Interleaving is frequently used in digital communication and storage systems to improve the performance of forward error correcting codes. Many communication channels are not memoryless: errors typically occur in bursts rather than independently. If the number of errors within a code word exceeds the error-correcting code's capability, it fails to recover the original code word. Interleaving ameliorates this problem by shuffling source symbols across several code words, thereby creating a more uniform distribution of errors. Therefore, interleaving is widely used for burst error-correction.

The analysis of modern iterated codes, like turbo codes and LDPC codes, typically assumes an independent distribution of errors. Systems using LDPC codes therefore typically employ additional interleaving across the symbols within a code word.

For turbo codes, an interleaver is an integral component and its proper design is crucial for good performance. The iterative decoding algorithm works best when there are not short cycles in the factor graph that represents the decoder; the interleaver is chosen to avoid short cycles.

Interleaver designs include:

- Rectangular (or uniform) interleavers.

- Convolutional interleavers.

- Random interleavers (where the interleaver is a known random permutation).

- S-random interleaver (where the interleaver is a known random permutation with the constraint that no input symbols within distance S appear within a distance of S in the output).

- Another possible construction is a contention-free quadratic permutation polynomial (QPP). It is used for example in the 3GPP Long Term Evolution mobile telecommunication standard.

In multi-carrier communication systems, interleaving across carriers may be employed to provide frequency diversity, e.g., to mitigate frequency-selective fading or narrowband interference.

Example:

Transmission without Interleaving

```
Error-free message:                    aaaabbbbccccddddeeeeffffgggg

Transmission with a burst error:       aaaabbbbccc____deeeeffffgggg
```

Here, each group of the same letter represents a 4-bit one-bit error-correcting codeword. The codeword cccc is altered in one bit and can be corrected, but the codeword dddd is altered in three bits, so either it cannot be decoded at all or it might be decoded incorrectly.

With Interleaving

```
Error-free code words:                    aaaabbbbccccddddeeeeffffgggg

Interleaved:                              abcdefgabcdefgabcdefgabcdefg

Transmission with a burst error:          abcdefgabcd____bcdefgabcdefg

Received code words after deinterleaving: aa_abbbbccccdddde_eef_ffg_gg
```

In each of the codewords "aaaa", "eeee", "ffff", and "gggg", only one bit is altered, so one-bit error-correcting code will decode everything correctly.

Transmission without Interleaving

```
Original transmitted sentence:         ThisIsAnExampleOfInterleaving

Received sentence with a burst error:  ThisIs_____pleOfInterleaving
```

The term "AnExample" ends up mostly unintelligible and difficult to correct.

With Interleaving

```
Transmitted sentence:                     ThisIsAnExampleOfInterleaving...

Error-free transmission:                  TIEpfeaghsxlIrv.iAaenli.snmOten.

Received sentence with a burst error:     TIEpfe_____Irv.iAaenli.snmOten.

Received sentence after deinterleaving:   T_isI_AnE_amp_eOfInterle_vin_...
```

No word is completely lost and the missing letters can be recovered with minimal guesswork.

Disadvantages of Interleaving

Use of interleaving techniques increases total delay. This is because the entire interleaved block must be received before the packets can be decoded. Also interleavers hide the structure of errors; without an interleaver, more advanced decoding algorithms can take advantage of the error structure and achieve more reliable communication than a simpler decoder combined with an interleaver.

Software for Error-correcting Codes

Simulating the behaviour of error-correcting codes (ECCs) in software is a common practice to design, validate and improve ECCs. The upcoming wireless 5G standard raises a new range of applications for the software ECCs: the Cloud Radio Access Networks (C-RAN) in a Software-defined radio (SDR) context. The idea is to directly use software ECCs in the communications. For instance in the 5G, the software ECCs could be located in the cloud and the antennas connected to this computing resources: improving this way the flexibility of the communication network and eventually increasing the energy efficiency of the system.

In this context, there are various available Open-source software listed below (non exhaustive).

- AFF3CT(A Fast Forward Error Correction Toolbox): a full communication chain in C++ (many supported codes like Turbo, LDPC, Polar codes, etc.), very fast and specialized on channel coding (can be used as a program for simulations or as a library for the SDR).

- IT++: a C++ library of classes and functions for linear algebra, numerical optimization, signal processing, communications, and statistics.

- OpenAir: implementation (in C) of the 3GPP specifications concerning the Evolved Packet Core Networks.

List of Error-correcting Codes

Distance	Code
2 (single-error detecting)	Parity
3 (single-error correcting)	Triple modular redundancy
3 (single-error correcting)	perfect Hamming such as Hamming(7,4)
4 (SECDED)	Extended Hamming
5 (double-error correcting)	
6 (double-error correct-/triple error detect)	
7 (three-error correcting)	perfect binary Golay code
8 (TECFED)	extended binary Golay code

- AN codes.

- BCH code, which can be designed to correct any arbitrary number of errors per code block.

- Berger code.

- Constant-weight code.

- Convolutional code.
- Expander codes.
- Group codes.
- Golay codes, of which the Binary Golay code is of practical interest.
- Goppa code, used in the McEliece cryptosystem.
- Hadamard code.
- Hagelbarger code.
- Hamming code.
- Latin square based code for non-white noise (prevalent for example in broadband over powerlines).
- Lexicographic code.
- Long code.
- Low-density parity-check code, also known as Gallager code, as the archetype for sparse graph codes.
- LT code, which is a near-optimal rateless erasure correcting code (Fountain code).
- m of n codes.
- Online code, a near-optimal rateless erasure correcting code.
- Polar code (coding theory).
- Raptor code, a near-optimal rateless erasure correcting code.
- Reed–Solomon error correction.
- Reed–Muller code.
- Repeat-accumulate code.
- Repetition codes, such as Triple modular redundancy.
- Spinal code, a rateless, nonlinear code based on pseudo-random hash functions·
- Tornado code, a near-optimal erasure correcting code, and the precursor to Fountain codes.
- Turbo code.
- Walsh–Hadamard code.
- Cyclic redundancy checks (CRCs) can correct 1-bit errors for messages at most bits long for optimal generator polynomials of degree $2^{n-1} - 1$.

Forward Error Correction

In telecommunication, information theory, and coding theory, forward error correction (FEC) or channel coding is a technique used for controlling errors in data transmission over unreliable or noisy communication channels. The central idea is the sender encodes the message in a redundant way by using an error-correcting code (ECC).

The redundancy allows the receiver to detect a limited number of errors that may occur anywhere in the message, and often to correct these errors without re-transmission. FEC gives the receiver the ability to correct errors without needing a reverse channel to request re-transmission of data, but at the cost of a fixed, higher forward channel bandwidth. FEC is therefore applied in situations where re-transmissions are costly or impossible, such as one-way communication links and when transmitting to multiple receivers in multicast. For example, in the case of a satellite orbiting around Uranus, a re-transmission because of decoding errors can create a delay of 5 hours. FEC information is usually added to mass storage (magnetic, optical and solid state/flash based) devices to enable recovery of corrupted data, is widely used in modems, is used on systems where the primary memory is ECC memory and in broadcast situations, where the receiver does not have capabilities to request retransmission or doing so would induce significant latency.

FEC processing in a receiver may be applied to a digital bit stream or in the demodulation of a digitally modulated carrier. For the latter, FEC is an integral part of the initial analog-to-digital conversion in the receiver. The Viterbi decoder implements a soft-decision algorithm to demodulate digital data from an analog signal corrupted by noise. Many FEC coders can also generate a bit-error rate (BER) signal which can be used as feedback to fine-tune the analog receiving electronics.

The maximum fractions of errors or of missing bits that can be corrected is determined by the design of the ECC, so different forward error correcting codes are suitable for different conditions. In general, a stronger code induces more redundancy that needs to be transmitted using the available bandwidth, which reduces the effective bit-rate while improving the received effective signal-to-noise ratio. The noisy-channel coding theorem of Claude Shannon answers the question of how much bandwidth is left for data communication while using the most efficient code that turns the decoding error probability to zero. This establishes bounds on the theoretical maximum information transfer rate of a channel with some given base noise level. His proof is not constructive, and hence gives no insight of how to build a capacity achieving code. However, after years of research, some advanced FEC systems like polar code come very close to the theoretical maximum.

BANDWIDTH

Communication channels are classified as analog or digital. Bandwidth refers to the data throughput capacity of any communication channel. As bandwidth increases, more information per unit of time can pass through the channel. A simple analogy compares a communication channel to a water pipe. The larger the pipe, the more water can flow through it at a faster rate, just as a high capacity communication channel allows more data to flow at a higher rate than is possible with a lower capacity channel.

In addition to describing the capacity of a communication channel, the term "bandwidth" is frequently, and somewhat confusingly, applied to information transport requirements. For example, it might be specified that a broadcast signal requires a channel with a bandwidth of six MHz to transmit a television signal without loss or distortion. Bandwidth limitations arise from the physical properties of matter and energy. Every physical transmission medium has a finite bandwidth. The bandwidth of any given medium determines its communications efficiency for voice, data, graphics, or full motion video.

Widespread use of the Internet has increased public awareness of telecommunications bandwidth because both consumers and service providers are interested in optimizing the speed of Internet access and the speed with which web pages appear on computer screens.

Analog Signals

Natural signals such as those associated with voice, music, or vision, are analog in nature. Analog signals are represented by a sine wave , and analog channel capacities are measured in hertz (Hz) or cycles per second. Analog signals vary in amplitude (signal strength) or frequency (signal pitch or tone). Analog bandwidth is calculated by finding the difference between the minimum and maximum amplitudes or frequencies found on the particular communication channel.

For example, the bandwidth allocation of a telephone voice grade channel, which is classified as narrowband , is normally about 4,000 Hz, but the voice channel actually uses frequencies from 300 to 3,400 Hz, yielding a bandwidth that is 3,100 Hz wide. The additional space or guardbands on each side of the voice channel serve to prevent signal overlap with adjacent channels and are also used for transmitting call management information.

Digital Signals

Signals in computing environments are digital. Digital signals are described as discrete, or discontinuous, because they are transmitted in small, separate units called bits. Digital channel capacities are measured in either bits per second (bps) or signal changes per second, which is known as the baud rate. Although these terms are frequently used interchangeably, bits per second and baud rate are technically not the same. Baud rate is an actual measure of the number of signal changes that occur per second rather than the number of bits actually transmitted per second. Prefixes used in the measurement of data transmission speeds include kilo (thousands), mega (millions), giga (thousands of millions), and tera (thousands of giga). To describe digital transmission capabilities in bits per second, notations such as Kbps, Mbps, Gbps, and Tbps are common.

The telephone system has been in a gradual transition from an analog to a digital network. In order to transmit a digital signal over a conventional analog telephone line, a modem is needed to modulate the signal of the sender and demodulate the signal for the receiver. The term modem is an abbreviation of modulate-demodulate. Although the core capacity of the telephone network has experienced an explosion in available bandwidth, local access to homes and businesses, referred to as the local loop in the telephone network, frequently is limited to analog modem connections. Digital transmission is popular because it is a reliable, high-speed service that eliminates the need for modems.

Broadband Communications

Financial and other business activities, software downloads, video conferencing, and distance education have created a need for greater bandwidth. The term broadband is used to refer to hardware and media that can support a wide bandwidth. Coaxial cable and microwave transmission are classified as broadband. Coaxial cable, used for cable television, has a bandwidth of 500,000,000 Hz, or 500 megahertz, and microwave transmission has a bandwidth of 10,000 Hz.

The capacity potential of broadband devices is considerably greater than that of narrowband technology, resulting in greater data transmission speeds and faster download speeds, which are important to Internet users. Data transmission speeds range from a low of 14,400 bps on a low speed modem to more than ten gigabits per second on a fiber optic cable. On the assumption that 50,000 bits represents a page of data, it takes 3.5 seconds to transmit the page at 14,400 bps, but only 8/10 of a second at 64,000 bps. If a page of graphics contains one million bits per page, it takes more than a minute to transmit the page at 14,400 bps, compared to 16 seconds at 64 Kbps. Full motion video requires an enormous bandwidth of 12 Mbps.

Upload versus Download Bandwidth

Among Internet Service Providers (ISPs) and broadband cable or satellite links, there is considerable difference in upstream, or upload, bandwidth and downstream, or download, bandwidth. Upstream transmission occurs when one sends information to an ISP whereas downstream transmission occurs when information is received from an ISP. For example, a broadband cable modem connection might transmit upstream at one Mbps and downstream at ten Mbps.

Typical media used to connect to the Internet, along with upstream and downstream bandwidths include: T3 leased lines, T1 leased lines, cable modems, asymmetric digital subscribe lines (ADSLs), integrated services digital networks (ISDNs), and dial-up modems. As noted in Gary P. Schneider and James T. Perry's book Electronic Commerce, T3 leased lines provide the fastest speeds (44,700 kbps for both upstream and downstream speeds) while the rates for T1 leased lines are 1,544 kbps, ISDNs are 128 kbps, and dial-up modems are 56 kbps. ADSL upstream and downstream speeds are 640 and 9,000 kbps, respectively, while cable modem speeds are 768 kbps upstream and 10,000 kbps downstream.

Each of the connections has advantages and disadvantages. As the speed of the medium increases in the broadband media beginning with T1 lines, costs increase substantially. Although classified as broadband, cable modems are considered optimal in price and performance for the home user.

In communication system the message signal is transmitted through a channel also known as medium. There are different types of transmission mediums which offer different bandwidths.

The commonly used transmission media are wire (coaxial cable and fibre optics cable) and free space.

Coaxial Cable

Coaxial cable is commonly used wire medium, it has a bandwidth of 750 MHz.

Handling frequency for coaxial cable is 18 GHz.

Radio Waves

Communication through free space uses radio waves. It has a bandwidth from a few hundred kHz to a few GHz. This range of frequencies are divided and used for different services.

Service	Frequency Range	Used for
Standard AM broadcast	540 - 1600 kHz	AM Radios walkie-talkie
FM broadcast	88 - 108 MHz	FM Radios walkie-talkie
Television	54 - 72 MHz	VHF (very high frequencies) Audio
	76 - 88 MHz	TV Video
	174 - 216 MHz	UHF (ultra-high frequencies) Audio
	420 - 890 MHz	TV Video
Cellular Mobile Radio	896 - 901 MHz	Mobile to base station
	840 - 935 MHz	Base station to mobile
Satellite Communication	5.925 - 6.425 GHz	Uplink
	3.7 - 4.2 GHz	Downlink

Optical Fibre

Communication through optical fibres has a bandwidth from 1 THz to 1000 THz i.e. from micro-waves to ultraviolet.

Optical fibre is the most favoured medium for communication in the modern world.

NOISE

In electronics, noise is an unwanted disturbance in an electrical signal. Noise generated by electronic devices varies greatly as it is produced by several different effects.

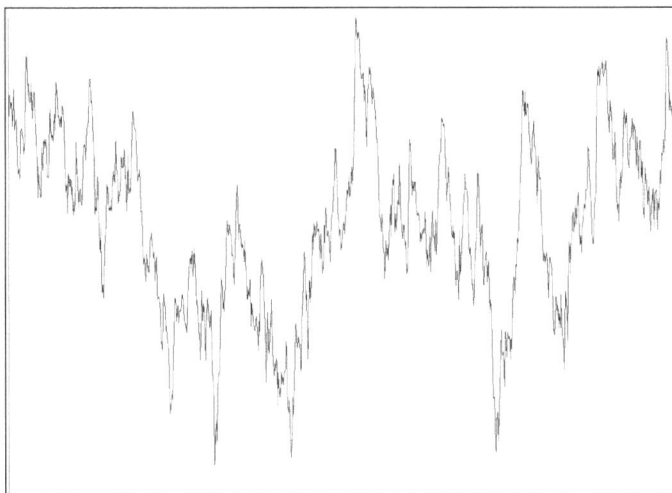

Analog display of random fluctuations in voltage in pink noise.

In communication systems, noise is an error or undesired random disturbance of a useful information signal. The noise is a summation of unwanted or disturbing energy from natural and sometimes man-made sources. Noise is, however, typically distinguished from interference, for example in the signal-to-noise ratio (SNR), signal-to-interference ratio (SIR) and signal-to-noise plus interference ratio (SNIR) measures. Noise is also typically distinguished from distortion, which is an unwanted systematic alteration of the signal waveform by the communication equipment, for example in signal-to-noise and distortion ratio (SINAD) and total harmonic distortion plus noise (THD+N) measures.

While noise is generally unwanted, it can serve a useful purpose in some applications, such as random number generation or dither.

Noise Types

Different types of noise are generated by different devices and different processes. Thermal noise is unavoidable at non-zero temperature while other types depend mostly on device type (such as shot noise, which needs a steep potential barrier) or manufacturing quality and semiconductor defects, such as conductance fluctuations, including 1/f noise.

Thermal Noise

Johnson–Nyquist noise (sometimes thermal, Johnson or Nyquist noise) is unavoidable, and generated by the random thermal motion of charge carriers (usually electrons), inside an electrical conductor, which happens regardless of any applied voltage.

Thermal noise is approximately white, meaning that its power spectral density is nearly equal throughout the frequency spectrum. The amplitude of the signal has very nearly a Gaussian probability density function. A communication system affected by thermal noise is often modeled as an additive white Gaussian noise (AWGN) channel.

Shot Noise

Shot noise in electronic devices results from unavoidable random statistical fluctuations of the electric current when the charge carriers (such as electrons) traverse a gap. If electrons flow across a barrier, then they have discrete arrival times. Those discrete arrivals exhibit shot noise. Typically, the barrier in a diode is used. Shot noise is similar to the noise created by rain falling on a tin roof. The flow of rain may be relatively constant, but the individual raindrops arrive discretely.

The root-mean-square value of the shot noise current i_n is given by the Schottky formula,

$$i_n = \sqrt{2Iq\Delta B}$$

where I is the DC current, q is the charge of an electron, and ΔB is the bandwidth in hertz. The Schottky formula assumes independent arrivals.

Vacuum tubes exhibit shot noise because the electrons randomly leave the cathode and arrive at the anode (plate). A tube may not exhibit the full shot noise effect: the presence of a space charge tends to smooth out the arrival times (and thus reduce the randomness of the current).

Conductors and resistors typically do not exhibit shot noise because the electrons thermalize and move diffusively within the material; the electrons do not have discrete arrival times. Shot noise has been demonstrated in mesoscopic resistors when the size of the resistive element becomes shorter than the electron–phonon scattering length.

Flicker Noise

Flicker noise, also known as $1/f$ noise, is a signal or process with a frequency spectrum that falls off steadily into the higher frequencies, with a pink spectrum. It occurs in almost all electronic devices and results from a variety of effects.

Burst Noise

Burst noise consists of sudden step-like transitions between two or more discrete voltage or current levels, as high as several hundred microvolts, at random and unpredictable times. Each shift in offset voltage or current lasts for several milliseconds to seconds. It is also known a popcorn noise for the popping or crackling sounds it produces in audio circuits.

Transit-time Noise

If the time taken by the electrons to travel from emitter to collector in a transistor becomes comparable to the period of the signal being amplified, that is, at frequencies above VHF and beyond, the transit-time effect takes place and noise input impedance of the transistor decreases. From the frequency at which this effect becomes significant, it increases with frequency and quickly dominates other sources of noise.

Coupled Noise

While noise may be generated in the electronic circuit itself, additional noise energy can be coupled into a circuit from the external environment, by inductive coupling or capacitive coupling, or through the antenna of a radio receiver.

Sources

- Intermodulation noise: Caused when signals of different frequencies share the same non-linear medium.

- Crosstalk: Phenomenon in which a signal transmitted in one circuit or channel of a transmission systems creates undesired interference onto a signal in another channel.

- Interference: Modification or disruption of a signal travelling along a medium.

- Atmospheric noise: This noise is also called static noise and it is the natural source of disturbance caused by lightning discharge in thunderstorm and the natural (electrical) disturbances occurring in nature.

- Industrial noise: Sources such as automobiles, aircraft, ignition electric motors and switching gear, High voltage wires and fluorescent lamps cause industrial noise. These noises are produced by the discharge present in all these operations.

- Solar noise: Noise that originates from the Sun is called solar noise. Under normal conditions there is constant radiation from the Sun due to its high temperature. Electrical disturbances such as corona discharges, as well as sunspots can produce additional noise. The intensity of solar noise varies over time in a solar cycle.

- Cosmic noise: Distant stars generate noise called cosmic noise. While these stars are too far away to individually affect terrestrial communications systems, their large number leads to appreciable collective effects. Cosmic noise has been observed in a range from 8 MHz to 1.43 GHz, the latter frequency corresponding to the 21-cm hydrogen line. Apart from man-made noise, it is the strongest component over the range of about 20 to 120 MHz. Little cosmic noise below 20MHz penetrates the ionosphere, while its eventual disappearance at frequencies in excess of 1.5 GHz is probably governed by the mechanisms generating it and its absorption by hydrogen in interstellar space.

Mitigation

In many cases noise found on a signal in a circuit is unwanted. There are many different noise reduction techniques that can reduce the noise picked up by a circuit.

- Faraday cage: A Faraday cage enclosing a circuit can be used to isolate the circuit from external noise sources. A faraday cage cannot address noise sources that originate in the circuit itself or those carried in on its inputs, including the power supply.

- Capacitive coupling: Capacitive coupling allows an AC signal from one part of the circuit to be picked up in another part through interaction of electric fields. Where coupling is unintended, the effects can be addressed through improved circuit layout and grounding.

- Ground loops: When grounding a circuit, it is important to avoid ground loops. Ground loops occur when there is a voltage difference between two ground connections. A good way to fix this is to bring all the ground wires to the same potential in a ground bus.

- Shielding cables: A shielded cable can be thought of as a Faraday cage for wiring and can protect the wires from unwanted noise in a sensitive circuit. The shield must be grounded to be effective. Grounding the shield at only one end can avoid a ground loop on the shield.

- Twisted pair wiring: Twisting wires in a circuit will reduce electromagnetic noise. Twisting the wires decreases the loop size in which a magnetic field can run through to produce a current between the wires. Small loops may exist between wires twisted together, but the magnetic field going through these loops induces a current flowing in opposite directions in alternate loops on each wire and so there is no net noise current.

- Notch filters: Notch filters or band-rejection filters are essential when eliminating a specific noise frequency. For example, in some countries power lines within a building run at 60 Hz. Sometimes a sensitive circuit will pick up this 60 Hz noise through some unwanted antenna (could be as simple as a wire in the circuit). Running the output through a notch filter at 60 Hz will amplify the desired signal without amplifying the 60 Hz noise. So in a sense the noise will be lost at the output of the filter.

Quantification

The noise level in an electronic system is typically measured as an electrical power N in watts or dBm, a root mean square (RMS) voltage (identical to the noise standard deviation) in volts, dBμV or a mean squared error (MSE) in volts squared. Noise may also be characterized by its probability distribution and noise spectral density $N_0(f)$ in watts per hertz.

A noise signal is typically considered as a linear addition to a useful information signal. Typical signal quality measures involving noise are signal-to-noise ratio (SNR or S/N), signal-to-quantization noise ratio (SQNR) in analog-to-digital conversion and compression, peak signal-to-noise ratio (PSNR) in image and video coding, E_b/N_0 in digital transmission, carrier to noise ratio (CNR) before the detector in carrier-modulated systems, and noise figure in cascaded amplifiers.

Noise is a random process, characterized by stochastic properties such as its variance, distribution, and spectral density. The spectral distribution of noise can vary with frequency, so its power density is measured in watts per hertz (W/Hz). Since the power in a resistive element is proportional to the square of the voltage across it, noise voltage (density) can be described by taking the square root of the noise power density, resulting in volts per root hertz ($V/\sqrt{H_z} V$). Integrated circuit devices, such as operational amplifiers commonly quote equivalent input noise level in these terms (at room temperature).

Noise power is measured in watts or decibels (dB) relative to a standard power, usually indicated by adding a suffix after dB. Examples of electrical noise-level measurement units are dBu, dBm0, dBrn, dBrnC, and dBrn($f_1 - f_2$), dBrn(144-line).

Noise levels are usually viewed in opposition to signal levels and so are often seen as part of a signal-to-noise ratio (SNR). Telecommunication systems strive to increase the ratio of signal level to noise level in order to effectively transmit data. In practice, if the transmitted signal falls below the level of the noise (often designated as the noise floor) in the system, data can no longer be decoded at the receiver. Noise in telecommunication systems is a product of both internal and external sources to the system.

In a carrier-modulated passband analog communication system, a certain carrier-to-noise ratio (CNR) at the radio receiver input would result in a certain signal-to-noise ratio in the detected message signal. In a digital communications system, a certain E_b/N_0 (normalized signal-to-noise ratio) would result in a certain bit error rate.

Dither

If the noise source is correlated with the signal, such as in the case of quantisation error, the intentional introduction of additional noise, called dither, can reduce overall noise in the bandwidth of interest. This technique allows retrieval of signals below the nominal detection threshold of an instrument. This is an example of stochastic resonance.

SIGNAL TO NOISE RATIO

Signal-to-noise ratio (abbreviated SNR or S/N) is a measure used in science and engineering that compares the level of a desired signal to the level of background noise. SNR is defined as the ratio of signal power to the noise power, often expressed in decibels. A ratio higher than 1:1 (greater than 0 dB) indicates more signal than noise.

While SNR is commonly quoted for electrical signals, it can be applied to any form of signal, for example isotope levels in an ice core, biochemical signaling between cells, or financial trading signals. Signal-to-noise ratio is sometimes used metaphorically to refer to the ratio of useful information to false or irrelevant data in a conversation or exchange. For example, in online discussion forums and other online communities, off-topic posts and spam are regarded as "noise" that interferes with the "signal" of appropriate discussion.

The signal-to-noise ratio, the bandwidth, and the channel capacity of a communication channel are connected by the Shannon–Hartley theorem.

Signal-to-noise ratio is defined as the ratio of the power of a signal to the power of background noise (unwanted signal):

$$\text{SNR} = \frac{P_\text{signal}}{P_\text{noise}},$$

where P is average power. Both signal and noise power must be measured at the same or equivalent points in a system, and within the same system bandwidth.

Depending on whether the signal is a constant (s) or a random variable (S), the signal to noise ratio for random noise N with expected value of zero becomes,

$$\text{SNR} = \frac{s^2}{\sigma_\text{N}^2}$$

or

$$\text{SNR} = \frac{E[S^2]}{\sigma_\text{N}^2}$$

where E refers to the expected value, i.e. in this case the mean of S^2.

If the signal and the noise are measured across the same impedance, the SNR can be obtained by calculating the square of the amplitude ratio,

$$\text{SNR} = \frac{P_\text{signal}}{P_\text{noise}} = \left(\frac{A_\text{signal}}{A_\text{noise}} \right)^2,$$

where A is root mean square (RMS) amplitude (for example, RMS voltage).

Decibels

Because many signals have a very wide dynamic range, signals are often expressed using the logarithmic decibel scale. Based upon the definition of decibel, signal and noise may be expressed in decibels (dB) as,

$$P_{\text{signal,dB}} = 10 \log_{10} \left(P_{\text{signal}} \right)$$

and

$$P_{\text{noise,dB}} = 10 \log_{10} \left(P_{\text{noise}} \right).$$

In a similar manner, SNR may be expressed in decibels as,

$$\text{SNR}_{\text{dB}} = 10 \log_{10} \left(\text{SNR} \right).$$

Using the definition of SNR,

$$\text{SNR}_{\text{dB}} = 10 \log_{10} \left(\frac{P_{\text{signal}}}{P_{\text{noise}}} \right).$$

Using the quotient rule for logarithms,

$$10 \log_{10} \left(\frac{P_{\text{signal}}}{P_{\text{noise}}} \right) = 10 \log_{10} \left(P_{\text{signal}} \right) - 10 \log_{10} \left(P_{\text{noise}} \right).$$

Substituting the definitions of SNR, signal, and noise in decibels into the above equation results in an important formula for calculating the signal to noise ratio in decibels, when the signal and noise are also in decibels:

$$\text{SNR}_{\text{dB}} = P_{\text{signal,dB}} - P_{\text{noise,dB}}.$$

In the above formula, P is measured in units of power, such as watts (W) or milliwatts (mW), and the signal-to-noise ratio is a pure number.

However, when the signal and noise are measured in volts (V) or amperes (A), which are measures of amplitude, they must first be squared to obtain a quantity proportional to power, as shown below:

$$\text{SNR}_{\text{dB}} = 10 \log_{10} \left[\left(\frac{A_{\text{signal}}}{A_{\text{noise}}} \right)^2 \right] = 20 \log_{10} \left(\frac{A_{\text{signal}}}{A_{\text{noise}}} \right) = \left(A_{\text{signal,dB}} - A_{\text{noise,dB}} \right).$$

Dynamic Range

The concepts of signal-to-noise ratio and dynamic range are closely related. Dynamic range measures the ratio between the strongest un-distorted signal on a channel and the minimum discernible signal, which for most purposes is the noise level. SNR measures the ratio between an arbitrary signal level (not necessarily the most powerful signal possible) and noise. Measuring signal-to-noise ratios requires the selection of a representative or reference signal. In audio engineering, the reference signal is usually a sine wave at a standardized nominal or alignment level, such as 1 kHz at +4 dBu (1.228 V_{RMS}).

SNR is usually taken to indicate an average signal-to-noise ratio, as it is possible that (near) instantaneous signal-to-noise ratios will be considerably different. The concept can be understood as normalizing the noise level to 1 (0 dB) and measuring how far the signal 'stands out'.

Difference from Conventional Power

In physics, the average power of an AC signal is defined as the average value of voltage times current; for resistive (non-reactive) circuits, where voltage and current are in phase, this is equivalent to the product of the rms voltage and current:

$$P = V_{rms} I_{rms}$$

$$P = \frac{V_{rms}^2}{R} = I_{rms}^2 R$$

But in signal processing and communication, one usually assumes that $R = 1\Omega$ so that factor is usually not included while measuring power or energy of a signal. This may cause some confusion among readers, but the resistance factor is not significant for typical operations performed in signal processing, or for computing power ratios. For most cases, the power of a signal would be considered to be simply,

$$P = V_{rms}^2 = \frac{A^2}{2}$$

where 'A' is the amplitude of the AC signal.

Alternative Definition

An alternative definition of SNR is as the reciprocal of the coefficient of variation, i.e., the ratio of mean to standard deviation of a signal or measurement,

$$SNR = \frac{\mu}{\sigma}$$

where μ is the signal mean or expected value and σ is the standard deviation of the noise, or an estimate thereof. Notice that such an alternative definition is only useful for variables that are always non-negative (such as photon counts and luminance). It is commonly used in image processing, where the SNR of an image is usually calculated as the ratio of the mean pixel value to the standard deviation of the pixel values over a given neighborhood. Sometimes SNR is defined as the square of the alternative definition above.

It should be also noted, that this definition is closely related to the Sensitivity Index or d', when assuming that the signal has two states, and the noise does not change between the two states.

The Rose criterion (named after Albert Rose) states that an SNR of at least 5 is needed to be able to distinguish image features at 100% certainty. An SNR less than 5 means less than 100% certainty in identifying image details.

Related measures are the "contrast ratio" and the "contrast-to-noise ratio".

SNR for Various Modulation Systems

Amplitude Modulation

Channel signal-to-noise ratio is given by,

$$(\text{SNR})_{C,AM} = \frac{A_C^2(1 + k_a^2 P)}{2WN_0}$$

where W is the bandwidth and k_a is modulation index,

Output signal-to-noise ratio (of AM receiver) is given by,

$$(\text{SNR})_{O,AM} = \frac{A_c^2 k_a^2 P}{2WN_0}$$

Frequency Modulation

Channel signal-to-noise ratio is given by,

$$(\text{SNR})_{C,FM} = \frac{A_c^2}{2WN_0}$$

Output signal-to-noise ratio is given by,

$$(\text{SNR})_{O,FM} = \frac{A_c^2 k_f^2 P}{2N_0 W^3}$$

Improving SNR in Practice

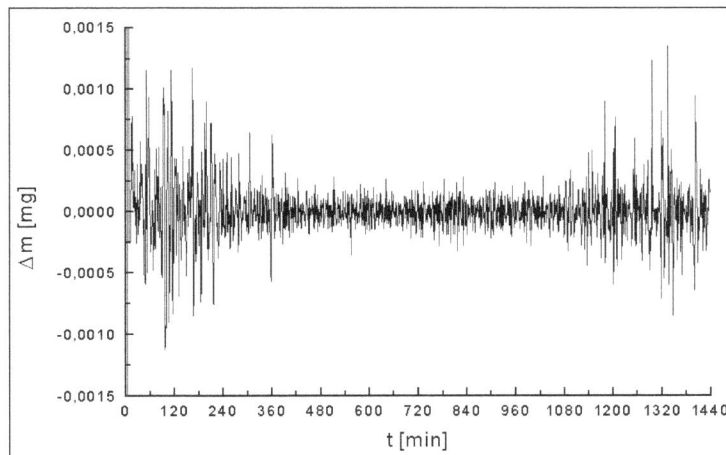

Recording of the noise of a thermogravimetric analysis device that is poorly isolated
from a mechanical point of view; the middle of the curve shows a lower noise,
due to a lesser surrounding human activity at night.

All real measurements are disturbed by noise. This includes electronic noise, but can also include external events that affect the measured phenomenon — wind, vibrations, gravitational attraction of the moon, variations of temperature, variations of humidity, etc., depending on what is measured and of the sensitivity of the device. It is often possible to reduce the noise by controlling the environment. Otherwise, when the characteristics of the noise are known and are different from the signals, it is possible to filter it or to process the signal.

For example, it is sometimes possible to use a lock-in amplifier to modulate and confine the signal within a very narrow bandwidth and then filter the detected signal to the narrow band where it resides, thereby eliminating most of the broadband noise.

When the signal is constant or periodic and the noise is random, it is possible to enhance the SNR by averaging the measurements. In this case the noise goes down as the square root of the number of averaged samples.

Additionally, internal noise of electronic systems can be reduced by low-noise amplifiers.

Digital Signals

When a measurement is digitized, the number of bits used to represent the measurement determines the maximum possible signal-to-noise ratio. This is because the minimum possible noise level is the error caused by the quantization of the signal, sometimes called quantization noise. This noise level is non-linear and signal-dependent; different calculations exist for different signal models. Quantization noise is modeled as an analog error signal summed with the signal before quantization ("additive noise").

This theoretical maximum SNR assumes a perfect input signal. If the input signal is already noisy (as is usually the case), the signal's noise may be larger than the quantization noise. Real analog-to-digital converters also have other sources of noise that further decrease the SNR compared to the theoretical maximum from the idealized quantization noise, including the intentional addition of dither.

Although noise levels in a digital system can be expressed using SNR, it is more common to use E_b/N_o, the energy per bit per noise power spectral density.

The modulation error ratio (MER) is a measure of the SNR in a digitally modulated signal.

Fixed Point

For n-bit integers with equal distance between quantization levels (uniform quantization) the dynamic range (DR) is also determined.

Assuming a uniform distribution of input signal values, the quantization noise is a uniformly distributed random signal with a peak-to-peak amplitude of one quantization level, making the amplitude ratio $2^n/1$. The formula is then:

$$\mathrm{DR}_{dB} = \mathrm{SNR}_{dB} = 20\log_{10}(2^n) \approx 6.02 \cdot n$$

This relationship is the origin of statements like "16-bit audio has a dynamic range of 96 dB". Each extra quantization bit increases the dynamic range by roughly 6 dB.

Assuming a full-scale sine wave signal (that is, the quantizer is designed such that it has the same minimum and maximum values as the input signal), the quantization noise approximates a sawtooth wave with peak-to-peak amplitude of one quantization level and uniform distribution. In this case, the SNR is approximately:

$$\text{SNR}_{dB} \approx 20\log_{10}(2^n \sqrt{3/2}) \approx 6.02 \cdot n + 1.761$$

Floating Point

Floating-point numbers provide a way to trade off signal-to-noise ratio for an increase in dynamic range. For n bit floating-point numbers, with n-m bits in the mantissa and m bits in the exponent:

$$\text{DR}_{dB} = 6.02 \cdot 2^m$$

$$\text{SNR}_{dB} = 6.02 \cdot (n - m)$$

Note that the dynamic range is much larger than fixed-point, but at a cost of a worse signal-to-noise ratio. This makes floating-point preferable in situations where the dynamic range is large or unpredictable. Fixed-point's simpler implementations can be used with no signal quality disadvantage in systems where dynamic range is less than 6.02m. The very large dynamic range of floating-point can be a disadvantage, since it requires more forethought in designing algorithms.

Optical SNR

Optical signals have a carrier frequency that is much higher than the modulation frequency (about 200 THz and more). This way the noise covers a bandwidth that is much wider than the signal itself. The resulting signal influence relies mainly on the filtering of the noise. To describe the signal quality without taking the receiver into account, the optical SNR (OSNR) is used. The OSNR is the ratio between the signal power and the noise power in a given bandwidth. Most commonly a reference bandwidth of 0.1 nm is used. This bandwidth is independent of the modulation format, the frequency and the receiver. For instance an OSNR of 20 dB/0.1 nm could be given, even the signal of 40 GBit DPSK would not fit in this bandwidth. OSNR is measured with an optical spectrum analyzer.

Types and Abbreviations

Signal to noise ratio may be abbreviated as SNR and less commonly as S/N. PSNR stands for Peak signal-to-noise ratio. GSNR stands for Geometric Signal-to-Noise Ratio. SINR is the Signal-to-noise-plus-interference ratio.

MODULATION ERROR RATIO

The modulation error ratio or MER is a measure used to quantify the performance of a digital radio (or digital TV) transmitter or receiver in a communications system using digital modulation (such as QAM). A signal sent by an ideal transmitter or received by a receiver would have all constellation points precisely at the ideal locations, however various imperfections in the implementation (such as noise, low image rejection ratio, phase noise, carrier suppression, distortion, etc.) or signal path cause the actual constellation points to deviate from the ideal locations.

Transmitter MER can be measured by specialised equipment, which demodulates the received signal in a similar way to how a real radio demodulator does it. Demodulated and detected signal can be used as a reasonably reliable estimate for the ideal transmitted signal in MER calculation.

An error vector is a vector in the I-Q plane between the ideal constellation point and the point received by the receiver. The Euclidean distance between the two points is its magnitude.

The modulation error ratio is equal to the ratio of the root mean square (RMS) power (in Watts) of the reference vector to the power (in Watts) of the error. It is defined in dB as,

$$\text{MER(dB)} = 10\log_{10}\left(\frac{P_{\text{signal}}}{P_{\text{error}}}\right)$$

where P_{error} is the RMS power of the error vector, and P_{signal} is the RMS power of ideal transmitted signal.

MER is defined as a percentage in a compatible (but reciprocal) way,

$$\text{MER(\%)} = \sqrt{\frac{P_{\text{error}}}{P_{\text{signal}}}} \times 100\%$$

with the same definitions.

MER is closely related to error vector magnitude (EVM), but MER is calculated from the average power of the signal. MER is also closely related to signal-to-noise ratio. MER includes all imperfections including deterministic amplitude imbalance, quadrature error and distortion, while noise is random by nature.

SHANNON–HARTLEY THEOREM

In information theory, the Shannon–Hartley theorem tells the maximum rate at which information can be transmitted over a communications channel of a specified bandwidth in the presence of noise. It is an application of the noisy-channel coding theorem to the archetypal case of a continuous-time analog communications channel subject to Gaussian noise. The theorem establishes Shannon's channel capacity for such a communication link, a bound on the maximum amount of error-free information per time unit that can be transmitted with a specified bandwidth in the presence of the noise interference, assuming that the signal power is bounded, and that the Gaussian noise process is characterized by a known power or power spectral density. The law is named after Claude Shannon and Ralph Hartley.

During the late 1920s, Harry Nyquist and Ralph Hartley developed a handful of fundamental ideas related to the transmission of information, particularly in the context of the telegraph as a communications system. At the time, these concepts were powerful breakthroughs individually, but they were not part of a comprehensive theory. In the 1940s, Claude Shannon developed the concept of channel capacity, based in part on the ideas of Nyquist and Hartley, and then formulated a complete theory of information and its transmission.

Statement of the Theorem

The Shannon–Hartley theorem states the channel capacity C, meaning the theoretical tightest upper bound on the information rate of data that can be communicated at an arbitrarily low error rate using an average received signal power S through an analog communication channel subject to additive white Gaussian noise of power N:

$$C = B \log_2 \left(1 + \frac{S}{N} \right)$$

where,

- c is the channel capacity in bits per second, a theoretical upper bound on the net bit rate (information rate, sometimes denoted I) excluding error-correction codes;

- B is the bandwidth of the channel in hertz (passband bandwidth in case of a bandpass signal);

- S is the average received signal power over the bandwidth (in case of a carrier-modulated passband transmission, often denoted C), measured in watts (or volts squared);

- N is the average power of the noise and interference over the bandwidth, measured in watts (or volts squared);

- S / N is the signal-to-noise ratio (SNR) or the carrier-to-noise ratio (CNR) of the communication signal to the noise and interference at the receiver (expressed as a linear power ratio, not as logarithmic decibels).

Implications of the Theorem

Comparison of Shannon's Capacity to Hartley's law

Comparing the channel capacity to the information rate from Hartley's law, we can find the effective number of distinguishable levels M: $2B \log_2 (M) = B \log_2 \left(1 + — \right)$.

$$M = \sqrt{1 + \frac{S}{N}}.$$

The square root effectively converts the power ratio back to a voltage ratio, so the number of levels is approximately proportional to the ratio of signal RMS amplitude to noise standard deviation.

This similarity in form between Shannon's capacity and Hartley's law should not be interpreted to mean that M pulse levels can be literally sent without any confusion. More levels are needed to allow for redundant coding and error correction, but the net data rate that can be approached with coding is equivalent to using that M in Hartley's law.

Frequency-dependent Case

In the simple version above, the signal and noise are fully uncorrelated, in which case $S + N$ is the total power of the received signal and noise together. A generalization of the above equation for the case where the additive noise is not white (or that the S / N is not constant with frequency over the bandwidth) is obtained by treating the channel as many narrow, independent Gaussian channels in parallel,

$$C = \int_0^B \log_2\left(1 + \frac{S(f)}{N(f)}\right) df$$

where,

- C is the channel capacity in bits per second.

- B is the bandwidth of the channel in Hz.

- $S(f)$ is the signal power spectrum.

- $N(f)$ is the noise power spectrum.

- f is frequency in Hz.

The theorem only applies to Gaussian stationary process noise. This formula's way of introducing frequency-dependent noise cannot describe all continuous-time noise processes. For example, consider a noise process consisting of adding a random wave whose amplitude is 1 or −1 at any point in time, and a channel that adds such a wave to the source signal. Such a wave's frequency components are highly dependent. Though such a noise may have a high power, it is fairly easy to transmit a continuous signal with much less power than one would need if the underlying noise was a sum of independent noises in each frequency band.

Approximations

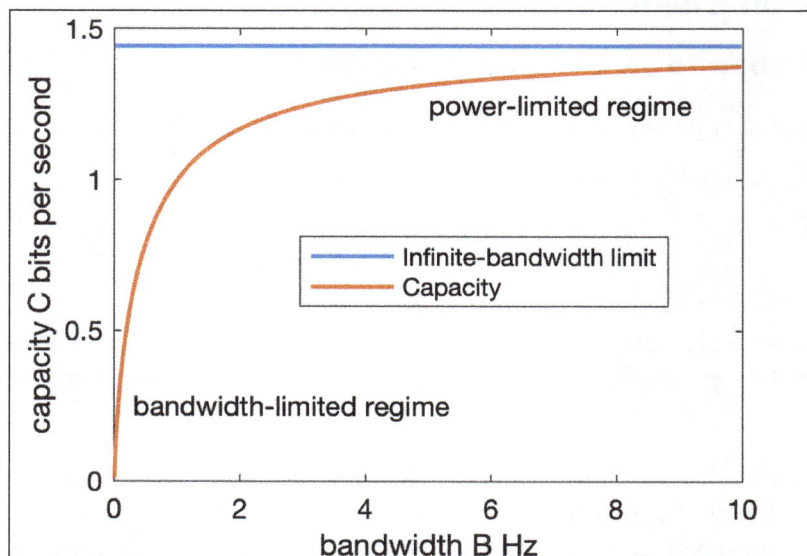

AWGN channel capacity with the power-limited regime and bandwidth-limited regime indicated.

Here,

$$\frac{S}{N_o} = 1,$$

B and C can be scaled proportionally for other values.

For large or small and constant signal-to-noise ratios.

Bandwidth-limited Case

When the SNR is large (S/N >> 1), the logarithm is approximated by,

$$\log_2\left(1+\frac{S}{N}\right) \approx \log_2\frac{S}{N} = \frac{\ln 10}{\ln 2}\cdot\log_{10}\frac{S}{N} \approx 3.32\cdot\log_{10}\frac{S}{N}$$

in which case the capacity is logarithmic in power and approximately linear in bandwidth (not quite linear, since N increases with bandwidth, imparting a logarithmic effect). This is called the bandwidth-limited regime.

$$C \approx 0.332 \cdot B \cdot \text{SNR (in dB)}$$

where,

$$\text{SNR (in dB)} = 10\log_{10}\frac{S}{N}.$$

Power-limited Case

Similarly, when the SNR is small (if S/N << 1), applying the approximation to the logarithm,

$$\log_2\left(1+\frac{S}{N}\right) = \frac{1}{\ln 2}\cdot\ln\left(1+\frac{S}{N}\right) \approx \frac{1}{\ln 2}\cdot\frac{S}{N} \approx 1.44\cdot\frac{S}{N}$$

then the capacity is linear in power and insensitive to bandwidth. This is called the power-limited regime.

$$C \approx 1.44 \cdot B \cdot \frac{S}{N}.$$

In this low-SNR approximation, capacity is independent of bandwidth if the noise is white, of spectral density N_0 watts per hertz, in which case the total noise power is $B \cdot N_0$.

$$C \approx 1.44 \cdot \frac{S}{N_0}$$

Examples:

1. At a SNR of 0 dB (Signal power = Noise power) the Capacity in bits/s is equal to the bandwidth in hertz.

2. If the SNR is 20 dB, and the bandwidth available is 4 kHz, which is appropriate for tele-phone communications, then C = 4000 $\log_2(1 + 100)$ = 4000 $\log_2$ (101) = 26.63 kbit/s. Note that the value of S/N = 100 is equivalent to the SNR of 20 dB.

3. If the requirement is to transmit at 50 kbit/s, and a bandwidth of 10 kHz is used, then the minimum S/N required is given by 50000 = 10000 $\log_2$(1+S/N) so C/B = 5 then S/N = 2^5 −1 = 31, corresponding to an SNR of 14.91 dB (10 x $\log_{10}$(31)).

4. What is the channel capacity for a signal having a 1 MHz bandwidth, received with a SNR of -30 dB ? That means a signal deeply buried in noise. -30 dB means a S/N = 10^{-3}. It leads to a maximal rate of information of $10^6 \log_2 (1 + 10^{-3})$ = 1443 bits/s. These values are typical of the received ranging signals of the GPS, where the navigation message is sent at 50 bits/s (below the channel capacity for the given S/N), and whose bandwidth is spread to around 1 MHz by a pseudo-noise multiplication before transmission.

5. Channel capacity is proportional to the bandwidth of the channel and to the logarithm of SNR. This means channel capacity can be increased linearly either by increasing the channel's bandwidth given a fixed SNR requirement or, with fixed bandwidth, by using higher-order modulations that need a very high SNR to operate. As the modulation rate increases, the spectral efficiency improves, but at the cost of the SNR requirement. Thus, there is an exponential rise in the SNR requirement if one adopts a 16QAM or 64QAM however, the spectral efficiency improves.

6. In MIMO, when the number of antenna beams are increased the channel capacity also gets increased. The correlation between the number of MIMO antennas and throughput is still not linear.

TRANSMITTER

In electronics and telecommunications a transmitter or radio transmitter is an electronic device which produces radio waves with an antenna. The transmitter itself generates a radio frequency alternating current, which is applied to the antenna. When excited by this alternating current, the antenna radiates radio waves.

Transmitters are necessary component parts of all electronic devices that communicate by radio, such as radio and television broadcasting stations, cell phones, walkie-talkies, wireless comput-er networks, Bluetooth enabled devices, garage door openers, two-way radios in aircraft, ships, spacecraft, radar sets and navigational beacons. The term transmitter is usually limited to equip-ment that generates radio waves for communication purposes; or radiolocation, such as radar and navigational transmitters. Generators of radio waves for heating or industrial purposes, such as microwave ovens or diathermy equipment, are not usually called transmitters, even though they often have similar circuits.

The term is popularly used more specifically to refer to a broadcast transmitter, a transmitter used in broadcasting, as in FM radio transmitter or television transmitter. This usage typically includes both the transmitter proper, the antenna, and often the building it is housed in.

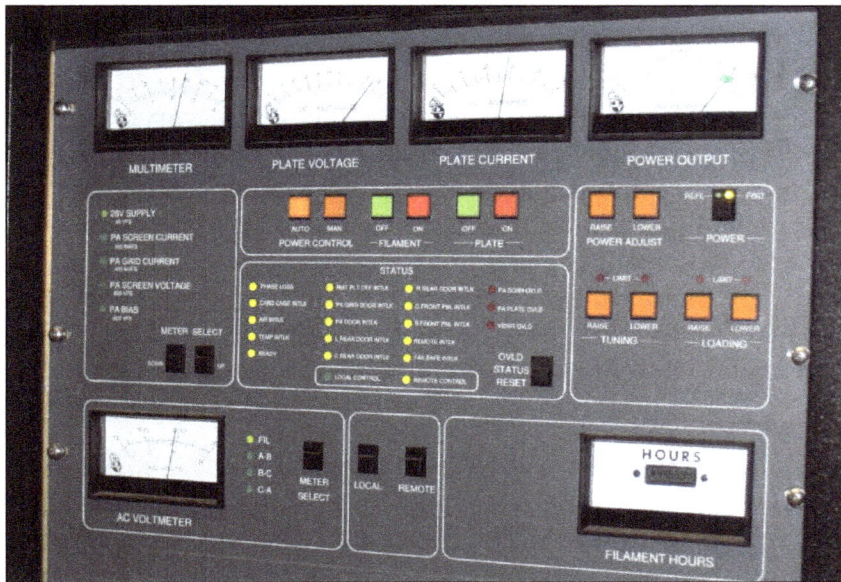

Commercial FM broadcasting transmitter at radio station WDET-FM.
It broadcasts at 101.9 MHz with a radiated power of 48 kW.

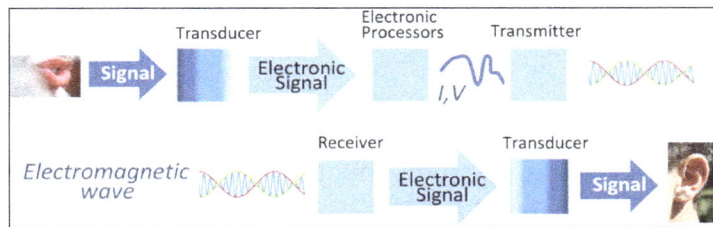

A radio transmitter is usually part of a radio communication system which uses electromagnetic
waves (radio waves) to transport information (in this case sound) over a distance.

A transmitter can be a separate piece of electronic equipment, or an electrical circuit within another electronic device. A transmitter and a receiver combined in one unit is called a transceiver. The term transmitter is often abbreviated "XMTR" or "TX" in technical documents. The purpose of most transmitters is radio communication of information over a distance. The information is provided to the transmitter in the form of an electronic signal, such as an audio (sound) signal from a microphone, a video (TV) signal from a video camera, or in wireless networking devices, a digital signal from a computer. The transmitter combines the information signal to be carried with the radio frequency signal which generates the radio waves, which is called the carrier signal. This process is called modulation. The information can be added to the carrier in several different ways, in different types of transmitters. In an amplitude modulation (AM) transmitter, the information is added to the radio signal by varying its amplitude. In a frequency modulation (FM) transmitter, it is added by varying the radio signal's frequency slightly. Many other types of modulation are also used.

The radio signal from the transmitter is applied to the antenna, which radiates the energy as radio waves. The antenna may be enclosed inside the case or attached to the outside of the transmitter, as in portable devices such as cell phones, walkie-talkies, and garage door openers. In more powerful transmitters, the antenna may be located on top of a building or on a separate tower, and connected to the transmitter by a feed line, that is a transmission line.

Operation

Electromagnetic waves are radiated by electric charges undergoing acceleration. Radio waves, electromagnetic waves of radio frequency, are generated by time-varying electric currents, consisting of electrons flowing through a metal conductor called an antenna which are changing their velocity or direction and thus accelerating. An alternating current flowing back and forth in an antenna will create an oscillating magnetic field around the conductor. The alternating voltage will also charge the ends of the conductor alternately positive and negative, creating an oscillating electric field around the conductor. If the frequency of the oscillations is high enough, in the radio frequency range above about 20 kHz, the oscillating coupled electric and magnetic fields will radiate away from the antenna into space as an electromagnetic wave, a radio wave.

A radio transmitter is an electronic circuit which transforms electric power from a power source into a radio frequency alternating current to apply to the antenna, and the antenna radiates the energy from this current as radio waves. The transmitter also impresses information such as an audio or video signal onto the radio frequency current to be carried by the radio waves. When they strike the antenna of a radio receiver, the waves excite similar (but less powerful) radio frequency currents in it. The radio receiver extracts the information from the received waves.

Components

A practical radio transmitter usually consists of these parts:

- A power supply circuit to transform the input electrical power to the higher voltages needed to produce the required power output.

- An electronic oscillator circuit to generate the radio frequency signal. This usually generates a sine wave of constant amplitude called the carrier wave, because it serves to "carry" the information through space. In most modern transmitters, this is a crystal oscillator in which the frequency is precisely controlled by the vibrations of a quartz crystal. The frequency of the carrier wave is considered the frequency of the transmitter.

- A modulator circuit to add the information to be transmitted to the carrier wave produced by the oscillator. This is done by varying some aspect of the carrier wave. The information is provided to the transmitter either in the form of an audio signal, which represents sound, a video signal which represents moving images, or for data in the form of a binary digital signal which represents a sequence of bits, a bitstream. Different types of transmitters use different modulation methods to transmit information:

  ○ In an AM (amplitude modulation) transmitter the amplitude (strength) of the carrier wave is varied in proportion to the modulation signal.

  ○ In an FM (frequency modulation) transmitter the frequency of the carrier is varied by the modulation signal.

  ○ In an FSK (frequency-shift keying) transmitter, which transmits digital data, the frequency of the carrier is shifted between two frequencies which represent the two binary digits, 0 and 1.

Many other types of modulation are also used. In large transmitters the oscillator and modulator together are often referred to as the exciter.

- A radio frequency (RF) amplifier to increase the power of the signal, to increase the range of the radio waves.

- An impedance matching (antenna tuner) circuit to match the impedance of the transmitter to the impedance of the antenna (or the transmission line to the antenna), to transfer power efficiently to the antenna. If these impedances are not equal, it causes a condition called standing waves, in which the power is reflected back from the antenna toward the transmitter, wasting power and sometimes overheating the transmitter.

In higher frequency transmitters, in the UHF and microwave range, free running oscillators are unstable at the output frequency. Older designs used an oscillator at a lower frequency, which was multiplied by frequency multipliers to get a signal at the desired frequency. Modern designs more commonly use an oscillator at the operating frequency which is stabilized by phase locking to a very stable lower frequency reference, usually a crystal oscillator.

RECEIVER

In radio communications, a radio receiver, also known as a receiver, wireless or simply radio is an electronic device that receives radio waves and converts the information carried by them to a usable form. It is used with an antenna. The antenna intercepts radio waves (electromagnetic waves) and converts them to tiny alternating currents which are applied to the receiver, and the receiver extracts the desired information. The receiver uses electronic filters to separate the desired radio frequency signal from all the other signals picked up by the antenna, an electronic amplifier to increase the power of the signal for further processing, and finally recovers the desired information through demodulation.

A bedside clock radio that combines a radio receiver with an alarm clock.

The information produced by the receiver may be in the form of sound, moving images (television), or data. A radio receiver may be a separate piece of electronic equipment, or an electronic circuit within another device. Radio receivers are very widely used in modern technology, as components of communications, broadcasting, remote control, and wireless networking systems. In consumer electronics, the terms radio and radio receiver are often used specifically for receivers designed to reproduce sound transmitted by radio broadcasting stations, historically the first mass-market commercial radio application.

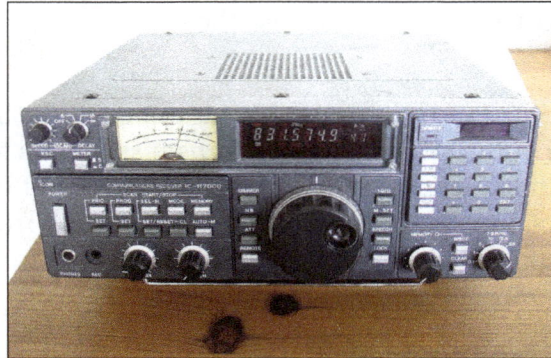

A modern communications receiver, used in two-way radio communication stations to talk with remote locations by shortwave radio.

Vacuum tube radio in the 1940s.

Broadcast Radio Receivers

The most familiar form of radio receiver is a broadcast receiver, often just called a radio, which receives audio programs intended for public reception transmitted by local radio stations. The sound is reproduced either by a loudspeaker in the radio or an earphone which plugs into a jack on the radio. The radio requires electric power, provided either by batteries inside the radio or a power cord which plugs into an electric outlet. All radios have a volume control to adjust the loudness of the audio, and some type of "tuning" control to select the radio station to be received.

Modulation Types

Modulation is the process of adding information to a radio carrier wave.

AM and FM

Two types of modulation are used in analog radio broadcasting systems; AM and FM.

In amplitude modulation (AM) the strength of the radio signal is varied by the audio signal. AM broadcasting is allowed in the AM broadcast bands which are between 148 and 283 kHz in the longwave range, and between 526 and 1706 kHz in the medium frequency (MF) range of the radio spectrum. AM broadcasting is also permitted in shortwave bands, between about 2.3 and 26 MHz, which are used for long distance international broadcasting.

In frequency modulation (FM) the frequency of the radio signal is varied slightly by the audio signal. FM broadcasting is permitted in the FM broadcast bands between about 65 and 108 MHz in the very high frequency (VHF) range. The exact frequency ranges vary somewhat in different countries.

FM stereo radio stations broadcast in stereophonic sound (stereo), transmitting two sound channels representing left and right microphones. A stereo receiver contains the additional circuits and parallel signal paths to reproduce the two separate channels. A monaural receiver, in contrast, only receives a single audio channel that is a combination (sum) of the left and right channels. While AM stereo transmitters and receivers exist, they have not achieved the popularity of FM stereo.

Most modern radios are "AM/FM" radios, and are able to receive both AM and FM radio stations, and have a switch to select which band to receive.

Digital Audio Broadcasting (DAB)

Digital audio broadcasting (DAB) is an advanced radio technology which debuted in some countries in 1998 that transmits audio from terrestrial radio stations as a digital signal rather than an analog signal as AM and FM do. Its advantages are that DAB has the potential to provide higher quality sound than FM (although many stations do not choose to transmit at such high quality), has greater immunity to radio noise and interference, makes better use of scarce radio spectrum bandwidth, and provides advanced user features such as electronic program guide, sports commentaries, and image slideshows. Its disadvantage is that it is incompatible with previous radios so that a new DAB receiver must be purchased. As of 2017, 38 countries offer DAB, with 2,100 stations serving listening areas containing 420 million people. Most countries plan an eventual switchover from FM to DAB. The United States and Canada have chosen not to implement DAB.

DAB radio stations work differently from AM or FM stations: a single DAB station transmits a wide 1,500 kHz bandwidth signal that carries from 9 to 12 channels from which the listener can choose. Broadcasters can transmit a channel at a range of different bit rates, so different channels can have different audio quality. In different countries DAB stations broadcast in either Band III (174–240 MHz) or L band (1.452–1.492 GHz).

Reception

The signal strength of radio waves decreases the farther they travel from the transmitter, so a radio station can only be received within a limited range of its transmitter. The range depends on the power of the transmitter, the sensitivity of the receiver, atmospheric and internal noise, as well as any geographical obstructions such as hills between transmitter and receiver. AM broadcast band radio waves travel as ground waves which follow the contour of the Earth, so AM radio stations can be reliably received at hundreds of miles distance. Due to their higher frequency, FM band radio signals cannot travel far beyond the visual horizon; limiting reception distance to about 40 miles (64 km), and can be blocked by hills between the transmitter and receiver. However FM radio is less susceptible to interference from radio noise (RFI, sferics, static) and has higher fidelity; better frequency response and less audio distortion, than AM. So in many countries serious music is only broadcast by FM stations, and AM stations specialize in radio news, talk radio, and sports. Like FM, DAB signals travel by line of sight so reception distances are limited by the visual horizon to about 30–40 miles (48–64 km).

Types of Broadcast Receiver

Radios are made in a range of styles and functions:

- Table radio - A self-contained radio with speaker designed to sit on a table.

- Clock radio - A bedside table radio that also includes an alarm clock. The alarm clock can be set to turn on the radio in the morning instead of an alarm, to wake the owner.

- Tuner - A high fidelity AM/FM radio receiver in a component home audio system. It has no speakers but outputs an audio signal which is fed into the system and played through the system's speakers.

- Portable radio - a radio powered by batteries that can be carried with a person. Radios are now often integrated with other audio sources in CD players and portable media players.

  o Boom box - a portable battery-powered high fidelity stereo sound system in the form of a box with a handle, which became popular during the mid 1970s.

  o Transistor radio - an older term for a portable pocket-sized broadcast radio receiver. Made possible by the invention of the transistor and developed in the 1950s, transistor radios were hugely popular during the 1960s and early 1970s, and changed the public's listening habits.

- Car radio - An AM/FM radio integrated into the dashboard of a vehicle, used for entertainment while driving. Virtually all modern cars and trucks are equipped with radios, which usually also includes a CD player.

- Satellite radio receiver - subscription radio receiver that receives audio programming from a direct broadcast satellite. The subscriber must pay a monthly fee. They are mostly designed as car radios.

- Shortwave receiver - This is a broadcast radio that also receives the shortwave bands. It is used for shortwave listening.

- AV receivers are a common component in a high-fidelity or home-theatre system; in addition to receiving radio programming, the receiver will also contain switching and amplifying functions to interconnect and control the other components of the system.

Other Applications

Radio receivers are essential components of all systems that use radio. Besides broadcast receivers, described above, radio receivers are used in a huge variety of electronic systems in modern technology. They can be a separate piece of equipment (a radio), or a subsystem incorporated into other electronic devices. A transceiver is a transmitter and receiver combined in one unit. Below is a list of a few of the most common types, organized by function:

- Broadcast television reception - Televisions receive a video signal representing a moving image, composed of a sequence of still images, and a synchronized audio signal representing the associated sound. The television channel received by a TV occupies a wider bandwidth than an audio signal, from 600 kHz to 6 MHz.

  ◦ Terrestrial television receiver, broadcast television or just television (TV) - Televisions contains an integral receiver (TV tuner) which receives free broadcast television from local television stations on TV channels in the VHF and UHF bands.

  ◦ Satellite TV receiver - a set-top box which receives subscription direct-broadcast satellite television, and displays it on an ordinary television. A rooftop satellite dish receives many channels all modulated on a K_u band microwave downlink signal from a geostationary direct broadcast satellite 22,000 miles (35,000 km) above the Earth, and the signal is converted to a lower intermediate frequency and transported to the box through a coaxial cable. The subscriber pays a monthly fee.

- Two-way voice communications - A two-way radio is an audio transceiver, a receiver and transmitter in the same device, used for bidirectional person-to-person voice communication. The radio link may be half-duplex, using a single radio channel in which only one radio can transmit at a time. so different users take turns talking, pressing a push to talk button on their radio which switches on the transmitter. Or the radio link may be full duplex, a bidirectional link using two radio channels so both people can talk at the same time, as in a cell phone.

  ◦ Cellphone - A portable telephone that is connected to the telephone network by radio signals exchanged with a local antenna called a cell tower. Cellphones have highly automated digital receivers in the UHF and microwave band that receive the incoming side of the duplex voice channel, as well as a control channel that handles dialing calls and switching the phone between cell towers. They usually also have several other receivers that connect them with other networks: a WiFi modem, a bluetooth modem, and a GPS receiver. The cell tower has sophisticated multichannel receivers that receive the signals from many cell phones simultaneously.

  ◦ Cordless phone - A landline telephone in which the handset is portable and communicates with the rest of the phone by a short range duplex radio link, instead of being attached by a cord. Both the handset and the base station have radio receivers operating in the UHF band that receive the short range bidirectional duplex radio link.

  ◦ Citizens band radio - A two-way half-duplex radio operating in the 27 MHz band that can be used without a license. They are often installed in vehicles and used by truckers and delivery services.

  ◦ Walkie-talkie - A handheld short range half-duplex two-way radio.

Handheld scanner.

○   Scanner - A receiver that continuously monitors multiple frequencies or radio channels by stepping through the channels repeatedly, listening briefly to each channel for a transmission. When a transmitter is found the receiver stops at that channel. Scanners are used to monitor emergency police, fire, and ambulance frequencies, as well as other two way radio frequencies such as citizens band. Scanning capabilities have also become a standard feature in communications receivers, walkie-talkies, and other two-way radios.

Modern communications receiver.

○   Communications receiver or shortwave receiver - A general purpose audio receiver covering the LF, MF, shortwave (HF), and VHF bands. Used mostly with a separate shortwave transmitter for two-way voice communication in communication stations, amateur radio stations, and for shortwave listening.

•   One-way (simplex) voice communications:

○   Wireless microphone receiver - These receive the short range signal from wireless microphones used onstage by musical artists, public speakers, and television personalities.

Baby monitor. The receiver is on the left.

- ◦ Baby monitor - This is a cribside appliance for mothers of infants that transmits the baby's sounds to a receiver carried by the mother, so she can monitor the baby while she is in other parts of the house. Many baby monitors now have video cameras to show a picture of the baby.

- Data communications:

  - ◦ Wireless (WiFi) modem - An automated short range digital data transmitter and receiver on a portable wireless device that communicates by microwaves with a nearby access point, a router or gateway, connecting the portable device with a local computer network (WLAN) to exchange data with other devices.

  - ◦ Bluetooth modem - a very short range (up to 10 m) 2.4-2.83 GHz data transceiver on a portable wireless device used as a substitute for a wire or cable connection, mainly to exchange files between portable devices and connect cellphones and music players with wireless earphones.

  - ◦ Microwave relay - a long distance high bandwidth point-to-point data transmission link consisting of a dish antenna and transmitter that transmits a beam of microwaves to another dish antenna and receiver. Since the antennas must be in line-of-sight, distances are limited by the visual horizon to 30–40 miles. Microwave links are used for private business data, wide area computer networks (WANs), and by telephone companies to transmit distance phone calls and television signals between cities.

- Satellite communications - Communication satellites are used for data transmission between widely separated points on Earth. Other satellites are used for search and rescue, remote sensing, weather reporting and scientific research. Radio communication with satellites and spacecraft can involve very long path lengths, from 35,786 km (22,236 mi) for geosynchronous satellites to billions of kilometers for interplanetary spacecraft. This and the limited power available to a spacecraft transmitter mean very sensitive receivers must be used.

  - ◦ Satellite transponder - A receiver and transmitter in a communications satellite that receives multiple data channels carrying long distance telephone calls, television signals. or internet traffic on a microwave uplink signal from a satellite ground station and retransmits the data to another ground station on a different downlink frequency. In a direct broadcast satellite the transponder broadcasts a stronger signal directly to satellite radio or satellite television receivers in consumer's homes.

  - ◦ Satellite ground station receiver - communication satellite ground stations receive data from communications satellites orbiting the Earth. Deep space ground stations such as those of the NASA Deep Space Network receive the weak signals from distant scientific spacecraft on interplanetary exploration missions. These have large dish antennas around 85 ft (25 m) in diameter, and extremely sensitive radio receivers similar to radio telescopes. The RF front end of the receiver is often cryogenically cooled to −195.79 °C (−320 °F) by liquid nitrogen to reduce radio noise in the circuit.

- Remote control - Remote control receivers receive digital commands that control a device, which may be as complex as a space vehicle or unmanned aerial vehicle, or as simple as a garage door opener. Remote control systems often also incorporate a telemetry channel to transmit data on the state of the controlled device back to the controller. Radio controlled model and other models include multichannel receivers in model cars, boats, airplanes, and helicopters. A short-range radio system is used in keyless entry systems.

- Radiolocation - This is the use of radio waves to determine the location or direction of an object.

  ○ Radar - A device that transmits a narrow beam of microwaves which reflect from a target back to a receiver, used to locate objects such as aircraft, spacecraft, missiles, ships or land vehicles. The reflected waves from the target are received by a receiver usually connected to the same antenna, indicating the direction to the target. Widely used in aviation, shipping, navigation, weather forecasting, space flight, vehicle collision avoidance systems, and the military.

  ○ Global navigation satellite system (GNSS) receiver, such as a GPS receiver used with the US Global Positioning System - the most widely used electronic navigation device. An automated digital receiver that receives simultaneous data signals from several satellites in low Earth orbit. Using extremely precise time signals it calculates the distance to the satellites, and from this the receiver's location on Earth. GNSS receivers are sold as portable devices, and are also incorporated in cell phones, vehicles and weapons, even artillery shells.

  ○ VOR receiver - Navigational instrument on an aircraft that uses the VHF signal from VOR navigational beacons between 108 and 117.95 MHz to determine the direction to the beacon very accurately, for air navigation.

  ○ Wild animal tracking receiver - A receiver with a directional antenna used to track wild animals which have been tagged with a small VHF transmitter, for wildlife management purposes.

- Other

  ○ Telemetry receiver - This receives data signals to monitor conditions of a process. Telemetry is used to monitor missile and spacecraft in flight, well logging during oil and gas drilling, and unmanned scientific instruments in remote locations.

  ○ Measuring receiver - A calibrated, laboratory grade radio receiver used to measure the characteristics of radio signals. Often incorporates a spectrum analyzer.

  ○ Radio telescope - Specialized antenna and radio receiver used as a scientific instrument to study weak radio waves from astronomical radio sources in space like stars, nebulas and galaxies in radio astronomy. They are the most sensitive radio receivers that exist, having large parabolic (dish) antennas up to 500 meters in diameter, and extremely sensitive radio circuits. The RF front end of the receiver is often cryogenically cooled by liquid nitrogen to reduce radio noise.

Filtering, Amplification and Demodulation

Practical radio receivers perform three basic functions on the signal from the antenna: filtering, amplification, and demodulation.

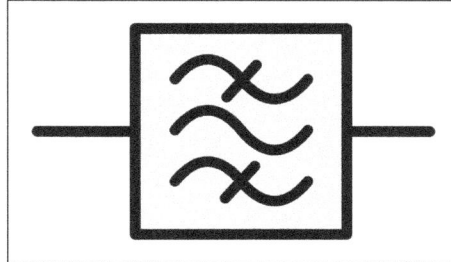

Symbol for a bandpass filter used in block diagrams of radio receivers.

- Bandpass filtering: Radio waves from many transmitters pass through the air simultaneously without interfering with each other. These can be separated in the receiver because they have different frequencies; that is, the radio wave from each transmitter oscillates at a different rate. To separate out the desired radio signal, the bandpass filter allows the frequency of the desired radio transmission to pass through, and blocks signals at all other frequencies.

  The bandpass filter consists of one or more resonant circuits (tuned circuits). The resonant circuit is connected between the antenna input and ground. When the incoming radio signal is at the resonant frequency, the resonant circuit has high impedance and the radio signal from the desired station is passed on to the following stages of the receiver. At all other frequencies the resonant circuit has low impedance, so signals at these frequencies are conducted to ground.

  - Bandwidth and selectivity: The information (modulation) in a radio transmission is contained in two narrow bands of frequencies called sidebands (SB) on either side of the carrier frequency (C), so the filter has to pass a band of frequencies, not just a single frequency. The band of frequencies received by the receiver is called its passband (PB), and the width of the passband in kilohertz is called the bandwidth (BW). The bandwidth of the filter must be wide enough to allow the sidebands through without distortion, but narrow enough to block any interfering transmissions on adjacent frequencies (such as S2 in the diagram). The ability of the receiver to reject unwanted radio stations near in frequency to the desired station is an important parameter called selectivity determined by the filter. In modern receivers quartz crystal, ceramic resonator, or surface acoustic wave (SAW) filters are often used which have sharper selectivity compared to networks of capacitor-inductor tuned circuits.

  - Tuning: To select a particular station the radio is "tuned" to the frequency of the desired transmitter. The radio has a dial or digital display showing the frequency it is tuned to. Tuning is adjusting the frequency of the receiver's passband to the frequency of the desired radio transmitter. Turning the tuning knob changes the resonant frequency of the tuned circuit. When the resonant frequency is equal to the radio transmitter's frequency the tuned circuit oscillates in sympathy, passing the signal on to the rest of the receiver.

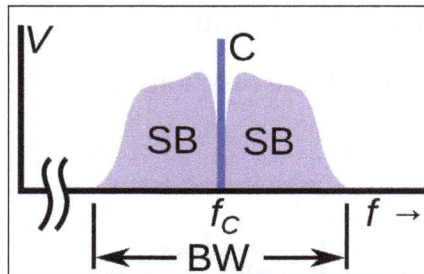

The frequency spectrum of a typical radio signal from an AM or FM radio transmitter. It consists of a strong component (C) at the carrier wave frequency f_C, with the modulation contained in narrow frequency bands called sidebands (SB) just above and below the carrier.

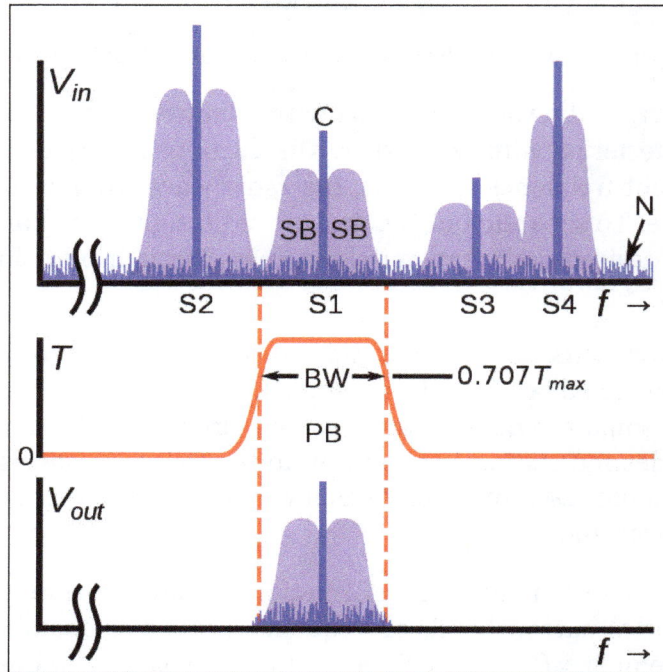

How the bandpass filter selects a single radio signal $S1$ from all the radio signals received by the antenna.

From top, the graphs show the voltage from the antenna applied to the filter V_{in}, the transfer function of the filter T, and the voltage at the output of the filter V_{out} as a function of frequency f. The transfer function T is the amount of signal that gets through the filter at each frequency.

$$V_{out}(f) = T(f)V_{in}(f)$$

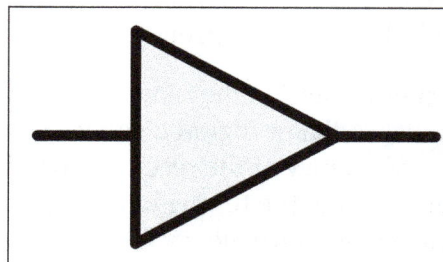

Symbol for an amplifier.

- Amplification: The power of the radio waves picked up by a receiving antenna decreases with the square of its distance from the transmitting antenna. Even with the powerful transmitters used in radio broadcasting stations, if the receiver is more than a few miles from the transmitter the power intercepted by the receiver's antenna is very small, perhaps as low as picowatts. To increase the power of the recovered signal, an amplifier circuit uses electric power from batteries or the wall plug to increase the amplitude (voltage or current) of the signal. In most modern receivers, the electronic components which do the actual amplifying are transistors.

Receivers usually have several stages of amplification: the radio signal from the bandpass filter is amplified to make it powerful enough to drive the demodulator, then the audio signal from the demodulator is amplified to make it powerful enough to operate the speaker. The degree of amplification of a radio receiver is measured by a parameter called its sensitivity, which is the minimum signal strength of a station at the antenna, measured in microvolts, necessary to receive the signal clearly, with a certain signal-to-noise ratio. Since it is easy to amplify a signal to any desired degree, the limit to the sensitivity of many modern receivers is not the degree of amplification but random electronic noise present in the circuit, which can drown out a weak radio signal.

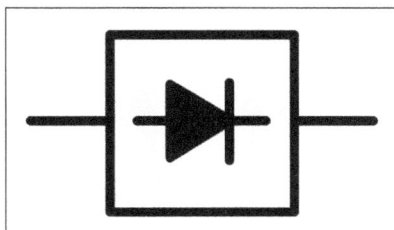

Symbol for a demodulator.

- Demodulation: After the radio signal is filtered and amplified, the receiver must extract the information-bearing modulation signal from the modulated radio frequency carrier wave. This is done by a circuit called a demodulator (detector). Each type of modulation requires a different type of demodulator:

  ○ An AM receiver that receives an (amplitude modulated) radio signal uses an AM demodulator.

  ○ An FM receiver that receives a frequency modulated signal uses an FM demodulator.

  ○ An FSK receiver which receives frequency shift keying (used to transmit digital data in wireless devices) uses an FSK demodulator.

Many other types of modulation are also used for specialized purposes.

The modulation signal output by the demodulator is usually amplified to increase its strength, then the information is converted back to a human-usable form by some type of transducer. An audio signal, representing sound, as in a broadcast radio, is converted to sound waves by an earphone or loudspeaker. A video signal, representing moving images, as in a television receiver, is converted to light by a display. Digital data, as in a wireless modem, is applied as input to a computer or microprocessor, which interacts with human users.

AM Demodulation

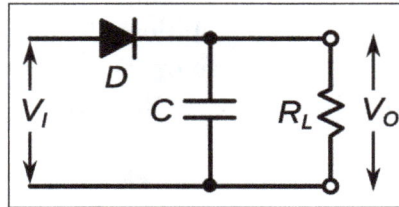

Envelope detector circuit.

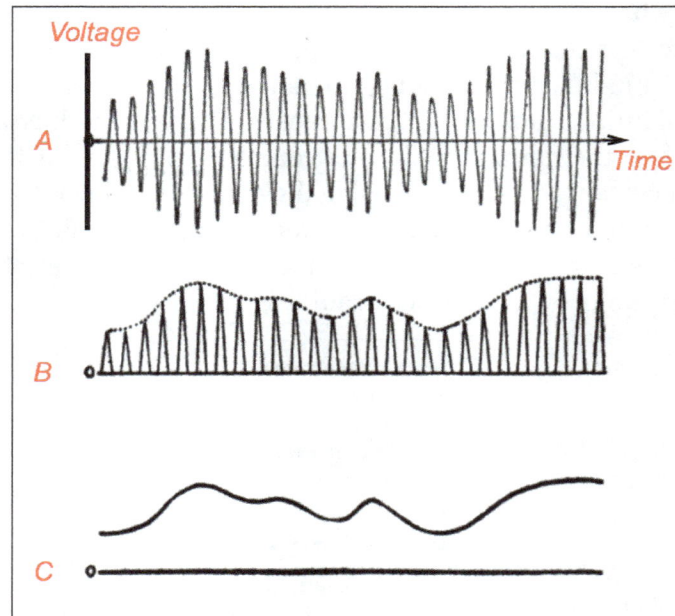

How an envelope detector works.

The easiest type of demodulation to understand is AM demodulation, used in AM radios to recover the audio modulation signal, which represents sound and is converted to sound waves by the radio's speaker. It is accomplished by a circuit called an envelope detector consisting of a diode (D) with a bypass capacitor (C) across its output.

The amplitude modulated radio signal from the tuned circuit is shown at (A). The rapid oscillations are the radio frequency carrier wave. The audio signal (the sound) is contained in the slow variations (modulation) of the amplitude (size) of the waves. If it was applied directly to the speaker, this signal cannot be converted to sound, because the audio excursions are the same on both sides of the axis, averaging out to zero, which would result in no net motion of the speaker's diaphragm. (B) When this signal is applied as input V_I to the detector, the diode (D) conducts current in one direction but not in the opposite direction, thus allowing through pulses of current on only one side of the signal. In other words, it rectifies the AC current to a pulsing DC current. The resulting voltage V_O applied to the load R_L no longer averages zero; its peak value is proportional to the audio signal. (C) The bypass capacitor (C) is charged up by the current pulses from the diode, and its voltage follows the peaks of the pulses, the envelope of the audio wave. It performs a smoothing (low pass filtering) function, removing the radio frequency carrier pulses, leaving the low frequency audio signal to pass through the load R_L. The audio signal is amplified and applied to earphones or a speaker.

Tuned Radio Frequency (TRF) Receiver

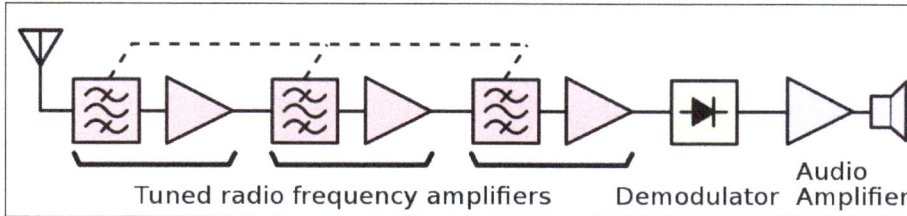

Block diagram of a tuned radio frequency receiver. To achieve enough selectivity to reject stations on adjacent frequencies, multiple cascaded bandpass filter stages had to be used. The dotted line indicates that the bandpass filters must be tuned together.

In the simplest type of radio receiver, called a tuned radio frequency (TRF) receiver, the three functions above are performed consecutively: (1) the mix of radio signals from the antenna is filtered to extract the signal of the desired transmitter; (2) this oscillating voltage is sent through a radio frequency (RF) amplifier to increase its strength to a level sufficient to drive the demodulator; (3) the demodulator recovers the modulation signal (which in broadcast receivers is an audio signal, a voltage oscillating at an audio frequency rate representing the sound waves) from the modulated radio carrier wave; (4) the modulation signal is amplified further in an audio amplifier, then is applied to a loudspeaker or earphone to convert it to sound waves.

Although the TRF receiver is used in a few applications, it has practical disadvantages which make it inferior to the superheterodyne receiver below, which is used in most applications. The drawbacks stem from the fact that in the TRF the filtering, amplification, and demodulation are done at the high frequency of the incoming radio signal. The bandwidth of a filter increases with its center frequency, so as the TRF receiver is tuned to different frequencies its bandwidth varies. Most important, the increasing congestion of the radio spectrum requires that radio channels be spaced very close together in frequency. It is extremely difficult to build filters operating at radio frequencies that have a narrow enough bandwidth to separate closely spaced radio stations. TRF receivers typically must have many cascaded tuning stages to achieve adequate selectivity. The Advantages section below describes how the superheterodyne receiver overcomes these problems.

The Superheterodyne design

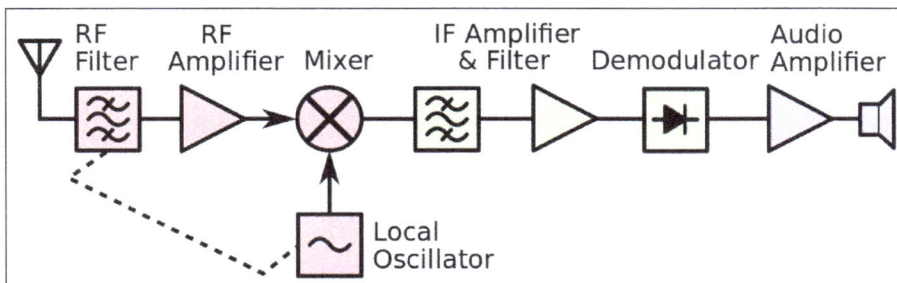

Block diagram of a superheterodyne receiver. The dotted line indicates that the RF filter and local oscillator must be tuned in tandem.

The superheterodyne receiver, invented in 1918 by Edwin Armstrong is the design used in almost all modern receivers except a few specialized applications.

In the superheterodyne, the radio frequency signal from the antenna is shifted down to a lower "intermediate frequency" (IF), before it is processed. The incoming radio frequency signal from the antenna is mixed with an unmodulated signal generated by a local oscillator (LO) in the receiver. The mixing is done in a nonlinear circuit called the "mixer". The result at the output of the mixer is a heterodyne or beat frequency at the difference between these two frequencies. The process is similar to the way two musical notes at different frequencies played together produce a beat note. This lower frequency is called the intermediate frequency (IF). The IF signal also has all the information that was present in the original RF signal. The IF signal passes through filter and amplifier stages, then is demodulated in a detector, recovering the original modulation.

The receiver is easy to tune; to receive a different frequency it is only necessary to change the local oscillator frequency. The stages of the receiver after the mixer operates at the fixed intermediate frequency (IF) so the IF bandpass filter does not have to be adjusted to different frequencies. The fixed frequency allows modern receivers to use sophisticated quartz crystal, ceramic resonator, or surface acoustic wave (SAW) IF filters that have very high Q factors, to improve selectivity.

The RF filter on the front end of the receiver is needed to prevent interference from any radio signals at the image frequency. Without an input filter the receiver can receive incoming RF signals at two different frequencies. The receiver can be designed to receive on either of these two frequencies; if the receiver is designed to receive on one, any other radio station or radio noise on the other frequency may pass through and interfere with the desired signal. A single tunable RF filter stage rejects the image frequency; since these are relatively far from the desired frequency, a simple filter provides adequate rejection. Rejection of interfering signals much closer in frequency to the desired signal is handled by the multiple sharply-tuned stages of the intermediate frequency amplifiers, which do no need to change their tuning. This filter does not need great selectivity, but as the receiver is tuned to different frequencies it must "track" in tandem with the local oscillator. The RF filter also serves to limit the bandwidth applied to the RF amplifier, preventing it from being overloaded by strong out-of-band signals.

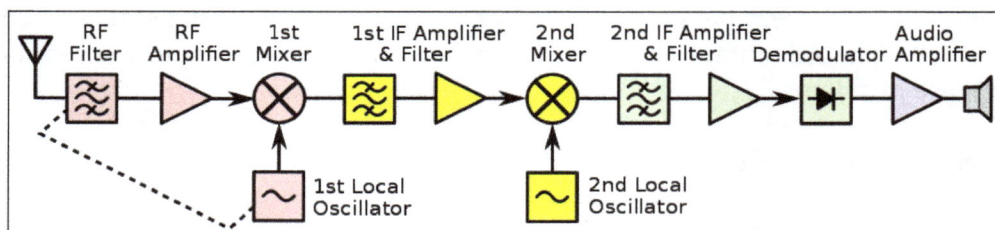

Block diagram of a dual-conversion superheterodyne receiver.

To achieve both good image rejection and selectivity, many modern superhet receivers use two intermediate frequencies; this is called a dual-conversion or double-conversion superheterodyne. The incoming RF signal is first mixed with one local oscillator signal in the first mixer to convert it to a high IF frequency, to allow efficient filtering out of the image frequency, then this first IF is mixed with a second local oscillator signal in a second mixer to convert it to a low IF frequency for good bandpass filtering. Some receivers even use triple-conversion.

At the cost of the extra stages, the superheterodyne receiver provides the advantage of greater selectivity than can be achieved with a TRF design. Where very high frequencies are in use, only the initial stage of the receiver needs to operate at the highest frequencies; the remaining stages can provide much of the receiver gain at lower frequencies which may be easier to manage. Tuning is simplified compared to a multi-stage TRF design, and only two stages need to track over the tuning range. The total amplification of the receiver is divided between three amplifiers at different frequencies; the RF, IF, and audio amplifier. This reduces problems with feedback and parasitic oscillations that are encountered in receivers where most of the amplifier stages operate at the same frequency, as in the TRF receiver.

The most important advantage is that better selectivity can be achieved by doing the filtering at the lower intermediate frequency. One of the most important parameters of a receiver is its bandwidth, the band of frequencies it accepts. In order to reject nearby interfering stations or noise, a narrow bandwidth is required. In all known filtering techniques, the bandwidth of the filter increases in proportion with the frequency, so by performing the filtering at the lower f_{IF} , rather than the frequency of the original radio signal f_{RF} , a narrower bandwidth can be achieved. Modern FM and television broadcasting, cellphones and other communications services, with their narrow channel widths, would be impossible without the superheterodyne.

Automatic Gain Control (AGC)

The signal strength (amplitude) of the radio signal from a receiver's antenna varies drastically, by orders of magnitude, depending on how far away the radio transmitter is, how powerful it is, and propagation conditions along the path of the radio waves. The strength of the signal received from a given transmitter varies with time due to changing propagation conditions of the path through which the radio wave passes, such as multipath interference; this is called fading. In an AM receiver the amplitude of the audio signal from the detector, and the sound volume, is proportional to the amplitude of the radio signal, so fading causes variations in the volume. In addition as the receiver is tuned between strong and weak stations, the volume of the sound from the speaker would vary drastically. Without an automatic system to handle it, in an AM receiver constant adjustment of the volume control would be required.

With other types of modulation like FM or FSK the amplitude of the modulation does not vary with the radio signal strength, but in all types the demodulator requires a certain range of signal amplitude to operate properly. Insufficient signal amplitude will cause an increase of noise in the demodulator, while excessive signal amplitude will cause amplifier stages to overload (saturate), causing distortion (clipping) of the signal.

Therefore, almost all modern receivers include a feedback control system which monitors the average level of the radio signal at the detector, and adjusts the gain of the amplifiers to give the optimum signal level for demodulation. This is called automatic gain control (AGC). AGC can be compared to the dark adaptation mechanism in the human eye; on entering a dark room the gain of the eye is increased by the iris opening. In its simplest form an AGC system consists of a rectifier which converts the RF signal to a varying DC level, a lowpass filter to smooth the variations and produce an average level. This is applied as a control signal to an earlier amplifier stage, to control its gain. In a superheterodyne receiver AGC is usually applied to the IF amplifier, and there may be a second AGC loop to control the gain of the RF amplifier to prevent it from overloading, too.

In certain receiver designs such as modern digital receivers, a related problem is DC offset of the signal. This is corrected by a similar feedback system.

References

- Pizzi, Skip; Jones, Graham (2014). A Broadcast Engineering Tutorial for Non-Engineers. CRC Press. P. 208. ISBN 978-1317906834

- "What's The Difference Between Bit Rate And Baud Rate?". Electronic Design. 2012-04-27. Retrieved 2018-01-18

- Huffman, William Cary; Pless, Vera S. (2003). Fundamentals of Error-Correcting Codes. Cambridge University Press. ISBN 978-0-521-78280-7

- "Documentation/edac.txt". Linux kernel documentation. Kernel.org. 2014-06-16. Archived from the original on 2009-09-05. Retrieved 2014-08-12

- Breeding, Andy (2004). The Music Internet Untangled: Using Online Services to Expand Your Musical Horizons. Giant Path. P. 128. ISBN 9781932340020

- Bandwidth-of-medium, communication: tutorialspoint.com, Retrieved 13 May, 2019

- Rafael C. González, Richard Eugene Woods (2008). Digital image processing. Prentice Hall. P. 354. ISBN 0-13-168728-X

- Bandwidth, astronomy-general, astronomy-and-space-exploration, science-and-technology: encyclopedia.com, Retrieved 6 January, 2019

- Anu A. Gokhale (2004). Introduction to Telecommunications (2nd ed.). Thomson Delmar Learning. ISBN 1-4018-5648-9

- Gary Cutlack (25 August 2010). "Mysterious Russian 'Numbers Station' Changes Broadcast After 20 Years". Gizmodo. Retrieved 12 March 2012

- Serway, Raymond; Faughn, Jerry; Vuille, Chris (2008). College Physics, 8th Ed. Cengage Learning. P. 714. ISBN 0495386936

Permissions

Index

A

Adaptive Delta Modulation, 26, 87
Aliasing, 4-5, 9, 15, 57-60, 62, 65, 74
Amplitude Fluctuations, 123-125
Amplitude Shift Keying, 35-36, 104, 108, 133
Analog Signals, 1-2, 28, 121, 195, 203
Analogue Cellular Phone, 158
Analogue-to-digital Conversion, 54
Audio Frequency-shift Keying, 113-114

B

Binary Phase-shift Keying, 116, 118, 161, 182

C

Channel Associated Signaling, 99
Channel Coding, 45, 104-106, 179, 185, 193, 200-202, 216
Channel Encoder, 3, 104
Constant Bit Rate, 173-176
Convolution Codes, 46-47

D

Delta Modulation, 22-24, 26, 66, 85-88
Differential Pulse Code Modulation, 73, 76, 82-84
Digital Modulation, 1, 4, 26, 35-38, 40, 66, 108, 115-116, 121, 128, 132, 144, 160, 165, 180, 215
Digital Multiplexing, 26-27, 94
Digital Signal Processing, 10, 15, 56, 62, 158
Direct Sequence Spread Spectrum, 49-50, 137
Discrete Memoryless Source, 43-44
Dpcm Transmitter, 21, 83

E

Encoding Speech, 27
Error Control Coding, 46

F

Fixed-length Code, 19-20
Fourier Transform, 8, 10, 27, 34, 40, 56, 58, 61, 110, 160
Frequency Division Multiplexing, 2, 28
Frequency Hopped Spread Spectrum, 49, 137
Frequency Shift Keying, 36, 104, 133, 233

G

Gaussian Filter, 113

Gaussian Function, 110-111
Gray Coding, 120, 124, 126-127, 129-130

H

Hamming Codes, 47, 189, 192, 195
Hamming Distance, 47, 127, 188, 196

I

Instant Message, 152

L

Least Significant Bits, 15, 76
Line Code, 28, 178-180, 182-185
Linear Predictive Coding, 26
Low Pass Filter, 4-5, 25, 60, 74, 77

M

M-ary Frequency Shift Keying, 133
Master Control Station, 164
Mid-tread Quantizers, 13
Minimum-shift Keying, 113, 118
Mobile Communication, 154, 157
Multiple Access Interference, 147

N

Narrow-band Signals, 48
Network Access Control, 156
Nyquist Rate, 7-8, 10, 22-23, 57, 65, 74, 82, 85, 181

O

Output Transducer, 3

P

Personal Communication Satellite Service, 163
Phase Shift Keying, 36-37, 104, 115, 118, 126, 128, 133
Plesiochronous Signals, 99, 101
Polar Signaling, 31
Power Spectral Density, 15, 33-35, 117, 206, 214, 216
Probability Density Function, 17, 110, 117, 206
Pulse Amplitude Modulation, 66-67, 74, 133
Pulse Code Modulation, 4, 66, 73-74, 76-77, 82-84
Pulse Modulation, 66, 83
Pulse Shaping, 106-107, 113
Pulse Width Modulation, 66, 69, 71

Q

Quadrature Amplitude Modulation, 38, 117-118, 126, 180

Quadrature Phase-shift Keying, 116, 120, 123, 160

Quantization Error, 10-11, 14-16, 20, 22, 75, 77-79, 84-85

Quantization Noise, 12, 15-17, 20, 78-81, 85, 209, 214

R

Radio Frequency, 73, 139, 150, 158, 220-223, 233-236

Reconstruction Filter, 6, 10

Remote Alarm Indication, 99

Role Based Access Control, 155

S

Scalar Quantizer, 17, 21

Signal-to-noise Ratio, 16, 40, 52, 85, 108, 122, 182, 185, 189, 194, 206, 209-217, 233

Source Encoder, 3, 13, 17, 44, 104

Spread Spectrum Technique, 2

Symbol Mapping, 106

Systematic Code, 46, 186

T

Time Division Multiplexing, 2, 88

Time-division Multiplexing, 88-89, 91, 93, 95, 139

U

Unipolar Signaling, 29-30

V

Variable Bit Rate, 173-175

Video Marketing, 151, 153

W

Wavelength Division Multiplexing, 28, 92

Wireless Application Protocol, 165

www.ingramcontent.com/pod-product-compliance
Lightning Source LLC
Chambersburg PA
CBHW061257190326

41458CB00011B/3701